AMAL
LUCAS

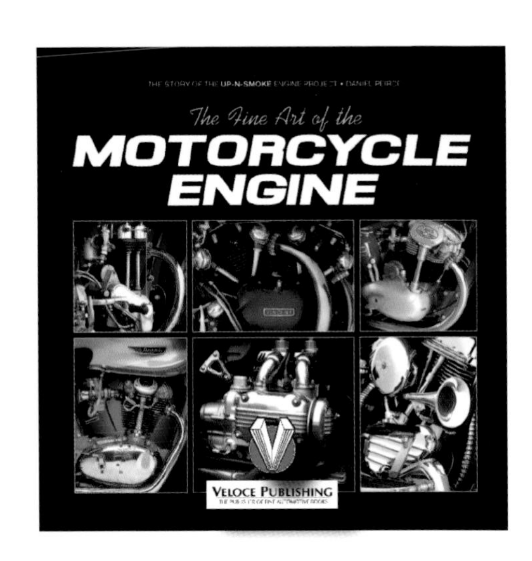
THE STORY OF THE UP-N-SMOKE ENGINE PROJECT • DANIEL PEIRCE
The Fine Art of the
MOTORCYCLE
ENGINE
VELOCE PUBLISHING

Other great books from Veloce –

Speedpro Series
4-cylinder Engine – How To Blueprint & Build A Short Block For High Performance (Hammill)
Alfa Romeo DOHC High-performance Manual (Kartalamakis)
Alfa Romeo V6 Engine High-performance Manual (Kartalamakis)
BMC 998cc A-series Engine – How To Power Tune (Hammill)
1275cc A-series High-performance Manual (Hammill)
Camshafts – How To Choose & Time Them For Maximum Power (Hammill)
Competition Car Datalogging Manual, The (Templeman)
Cylinder Heads – How To Build, Modify & Power Tune Updated & Revised Edition (Burgess & Gollan)
Distributor-type Ignition Systems – How To Build & Power Tune (Hammill)
Fast Road Car – How To Plan And Build Revised & Updated Colour New Edition (Stapleton)
Ford SOHC 'Pinto' & Sierra Cosworth DOHC Engines – How To Power Tune Updated & Enlarged Edition (Hammill)
Ford V8 – How To Power Tune Small Block Engines (Hammill)
Harley-Davidson Evolution Engines – How To Build & Power Tune (Hammill)
Holley Carburetors – How To Build & Power Tune Revised & Updated Edition (Hammill)
Jaguar XK Engines – How To Power Tune Revised & Updated Colour Edition (Hammill)
MG Midget & Austin-Healey Sprite – How To Power Tune Updated & Revised Edition (Stapleton)
MGB 4-cylinder Engine – How To Power Tune (Burgess)
MGB V8 Power – How To Give Your, Third Colour Edition (Williams)
MGB, MGC & MGB V8 – How To Improve (Williams)
Mini Engines – How To Power Tune On A Small Budget Colour Edition (Hammill)
Motorcycle-engined Racing Car – How To Build (Pashley)
Motorsport – Getting Started in (Collins)
Nitrous Oxide High-performance Manual, The (Langfield)
Rover V8 Engines – How To Power Tune (Hammill)
Sportscar/kitcar Suspension & Brakes – How To Build & Modify Enlarged & Updated 2nd Edition (Hammill)
SU Carburettor High-performance Manual (Hammill)
Suzuki 4x4 – How To Modify For Serious Off-road Action (Richardson)
Tiger Avon Sportscar – How To Build Your Own Updated & Revised 2nd Edition (Dudley)
TR2, 3 & TR4 – How To Improve (Williams)
TR5, 250 & TR6 – How To Improve (Williams)
TR7 & TR8 – How To Improve (Williams)
V8 Engine – How To Build A Short Block For High Performance (Hammill)
Volkswagen Beetle Suspension, Brakes & Chassis – How To Modify For High Performance (Hale)
Volkswagen Bus Suspension, Brakes & Chassis – How To Modify For High Performance (Hale)
Weber DCOE, & Dellorto DHLA Carburetors – How To Build & Power Tune 3rd Edition (Hammill)

Those Were The Days ... Series
Alpine Trials & Rallies 1910-1973 (Pfundner)
Austerity Motoring (Bobbitt)
Brighton National Speed Trials (Gardiner)
British Police Cars (Walker)
British Woodies (Peck)
Dune Buggy Phenomenon (Hale)
Dune Buggy Phenomenon Volume 2 (Hale)
Hot Rod & Stock Car Racing in Britain In The 1980s (Neil)
MG's Abingdon Factory (Moylan)
Motor Racing At Brands Hatch In The Seventies (Parker)
Motor Racing At Crystal Palace (Collins)
Motor Racing At Goodwood In The Sixties (Gardiner)
Motor Racing At Nassau In The 1950s & 1960s (O'Neil)
Motor Racing At Oulton Park In The 1960s (Mcfadyen)
Motor Racing At Oulton Park In The 1970s (Mcfadyen)
Three Wheelers (Bobbitt)

Enthusiast's Restoration Manual Series
Citroën 2CV, How To Restore (Porter)
Classic Car Bodywork, How To Restore (Thaddeus)
Classic Car Electrics (Thaddeus)
Classic Cars, How To Paint (Thaddeus)
Reliant Regal, How To Restore (Payne)
Triumph TR2/3/3A, How To Restore (Williams)
Triumph TR4/4A, How To Restore (Williams)
Triumph TR5/250 & 6, How To Restore (Williams)
Triumph TR7/8, How To Restore (Williams)
Volkswagen Beetle, How To Restore (Tyler)
VW Bay Window Bus (Paxton)
Yamaha FS1-E, How To Restore (Watts)

Essential Buyer's Guide Series
Alfa GT (Booker)
Alfa Romeo Spider Giulia (Booker & Talbott)
BMW GS (Henshaw)
BSA Bantam (Henshaw)
BSA Twins (Henshaw)
Citroën 2CV (Paxton)
Citroën ID & DS (Heilig)
Fiat 500 & 600 (Bobbitt)
Jaguar E-type 3.8 & 4.2-litre (Crespin)
Jaguar E-type V12 5.3-litre (Crespin)
Jaguar/Daimler XJ6, XJ12 & Sovereign (Crespin)
Jaguar XJ-S (Crespin)
MGB & MGB GT (Williams)
Mercedes-Benz 280SL-560DSL Roadsters (Bass)
Mercedes-Benz 'Pagoda' 230SL, 250SL & 280SL Roadsters & Coupés (Bass)
Morris Minor & 1000 (Newell)
Porsche 928 (Hemmings)
Rolls-Royce Silver Shadow & Bentley T-Series (Bobbitt)
Subaru Impreza (Hobbs)
Triumph Bonneville (Henshaw)
Triumph TR6 (Williams)
VW Beetle (Cservenka & Copping)
VW Bus (Cservenka & Copping)

Auto-Graphics Series
Fiat-based Abarths (Sparrow)
Jaguar MKI & II Saloons (Sparrow)
Lambretta Li Series Scooters (Sparrow)

Rally Giants Series
Audi Quattro (Robson)
Austin Healey 100-6 & 3000 (Robson)
Fiat 131 Abarth (Robson)
Ford Escort MkI (Robson)
Ford Escort RS Cosworth & World Rally Car (Robson)
Ford Escort RS1800 (Robson)
Lancia Stratos (Robson)
Peugeot 205 T16 (Robson)
Subaru Impreza (Robson)

General
1½-litre GP Racing 1961-1965 (Whitelock)
AC Two-litre Saloons & Buckland Sportscars (Archibald)
Alfa Romeo Giulia Coupé GT & GTA (Tipler)
Alfa Romeo Montreal – The Essential Companion (Taylor)
Alfa Tipo 33 (McDonough & Collins)
Alpine & Renault – The Development Of The Revolutionary Turbo F1 Car 1968 to 1979 (Smith)
Anatomy Of The Works Minis (Moylan)
Armstrong-Siddeley (Smith)
Autodrome (Collins & Ireland)
Automotive A-Z, Lane's Dictionary Of Automotive Terms (Lane)
Automotive Mascots (Kay & Springate)
Bahamas Speed Weeks, The (O'Neil)
Bentley Continental, Corniche And Azure (Bennett)
Bentley MkVI, Rolls-Royce Silver Wraith, Dawn & Cloud/Bentley R & S-Series (Nutland)
BMC Competitions Department Secrets (Turner, Chambers Browning)
BMW 5-Series (Cranswick)
BMW Z-Cars (Taylor)
Britains Farm Model Balers & Combines 1967 to 2007 (Pullen)
British 250cc Racing Motorcycles (Pereira)
British Cars, The Complete Catalogue Of, 1895-1975 (Culshaw & Horrobin)
BRM – A Mechanic's Tale (Salmon)
BRM V16 (Ludvigsen)
BSA Bantam Bible, The (Henshaw)
Bugatti Type 40 (Price)
Bugatti 46/50 Updated Edition (Price & Arbey)
Bugatti T44 & T49 (Price & Arbey)
Bugatti 57 2nd Edition (Price)
Caravans, The Illustrated History 1919-1959 (Jenkinson)
Caravans, The Illustrated History From 1960 (Jenkinson)
Carrera Panamericana, La (Tipler)
Chrysler 300 – America's Most Powerful Car 2nd Edition (Ackerson)
Chrysler PT Cruiser (Ackerson)
Citroën DS (Bobbitt)
Cliff Allison – From The Fells To Ferrari (Gauld)
Cobra – The Real Thing! (Legate)
Cortina – Ford's Bestseller (Robson)
Coventry Climax Racing Engines (Hammill)
Daimler SP250 New Edition (Long)
Datsun Fairlady Roadster To 280ZX – The Z-Car Story (Long)
Dino – The V6 Ferrari (Long)
Dodge Charger – Enduring Thunder (Ackerson)
Dodge Dynamite! (Grist)
Donington (Boddy)
Draw & Paint Cars – How To (Gardiner)
Drive On The Wild Side, A – 20 Extreme Driving Adventures From Around The World (Weaver)
Ducati 750 Bible, The (Falloon)
Ducati 860, 900 And Mille Bible, The (Falloon)
Dune Buggy, Building A – The Essential Manual (Shakespeare)
Dune Buggy Files (Hale)
Dune Buggy Handbook (Hale)
Edward Turner: The Man Behind The Motorcycles (Clew)
Fiat & Abarth 124 Spider & Coupé (Tipler)
Fiat & Abarth 500 & 600 2nd Edition (Bobbitt)
Fiats, Great Small (Ward)
Fine Art Of The Motorcycle Engine, The (Peirce)
Ford F100/F150 Pick-up 1948-1996 (Ackerson)
Ford F150 Pick-up 1997-2005 (Ackerson)
Ford GT – Then, And Now (Streather)
Ford GT40 (Legate)
Ford In Miniature (Olson)
Ford Model Y (Roberts)
Ford Thunderbird From 1954, The Book Of The (Long)
Forza Minardi! (Vigar)
Funky Mopeds (Skelton)
Gentleman Jack (Gauld)
GM In Miniature (Olson)
GT – The World's Best GT Cars 1953-73 (Dawson)
Hillclimbing & Sprinting – The Essential Manual (Short & Wilkinson)
Honda NSX (Long)
Jaguar, The Rise Of (Price)
Jaguar XJ-S (Long)
Jeep CJ (Ackerson)
Jeep Wrangler (Ackerson)
Karmann-Ghia Coupé & Convertible (Bobbitt)
Lamborghini Miura Bible, The (Sackey)
Lambretta Bible, The (Davies)
Lancia 037 (Collins)
Lancia Delta HF Integrale (Blaettel & Wagner)
Land Rover, The Half-ton Military (Cook)
Laverda Twins & Triples Bible 1968-1986 (Falloon)
Lea-Francis Story, The (Price)
Lexus Story, The (Long)
little book of smart, the (Jackson)
Lola – The Illustrated History (1957-1977) (Starkey)
Lola – All The Sports Racing & Single-seater Racing Cars 1978-1997 (Starkey)
Lola T70 – The Racing History & Individual Chassis Record 4th Edition (Starkey)
Lotus 49 (Oliver)
Marketingmobiles, The Wonderful Wacky World Of (Hale)
Mazda MX-5/Miata 1.6 Enthusiast's Workshop Manual (Grainger & Shoemark)
Mazda MX-5/Miata 1.8 Enthusiast's Workshop Manual (Grainger & Shoemark)
Mazda MX-5 Miata: The Book Of The World's Favourite Sportscar (Long)
Mazda MX-5 Miata Roadster (Long)
MGA (Price Williams)
MGB & MGB GT– Expert Guide (Auto-doc Series) (Williams)
MGB Electrical Systems (Astley)
Micro Caravans (Jenkinson)
Micro Trucks (Mort)
Microcars At Large! (Quellin)
Mini Cooper – The Real Thing! (Tipler)
Mitsubishi Lancer Evo, The Road Car & WRC Story (Long)
Montlhéry, The Story Of The Paris Autodrome (Boddy)
Morgan Maverick (Lawrence)
Morris Minor, 60 Years On The Road (Newell)
Moto Guzzi Sport & Le Mans Bible (Falloon)
Motor Movies – The Posters! (Veysey)
Motor Racing – Reflections Of A Lost Era (Carter)
Motorcycle Apprentice (Cakebread)
Motorcycle Road & Racing Chassis Designs (Noakes)
Motorhomes, The Illustrated History (Jenkinson)
Motorsport In colour, 1950s (Wainwright)
Nissan 300ZX & 350Z – The Z-Car Story (Long)
Off-Road Giants! – Heroes of 1960s Motorcycle Sport (Westlake)
Pass The Theory And Practical Driving Tests (Gibson & Hoole)
Peking To Paris 2007 (Young)
Plastic Toy Cars Of The 1950s & 1960s (Ralston)
Pontiac Firebird (Cranswick)
Porsche Boxster (Long)
Porsche 964, 993 & 996 Data Plate Code Breaker (Streather)
Porsche 356 (2nd Edition) (Long)
Porsche 911 Carrera – The Last Of The Evolution (Corlett)
Porsche 911R, RS & RSR, 4th Edition (Starkey)
Porsche 911 – The Definitive History 1963-1971 (Long)
Porsche 911 – The Definitive History 1971-1977 (Long)
Porsche 911 – The Definitive History 1977-1987 (Long)
Porsche 911 – The Definitive History 1987-1997 (Long)
Porsche 911 – The Definitive History 1997-2004 (Long)
Porsche 911SC 'Super Carrera' – The Essential Companion (Streather)
Porsche 914 & 914-6: The Definitive History Of The Road & Competition Cars (Long)
Porsche 924 (Long)
Porsche 944 (Long)
Porsche 993 'King Of Porsche' – The Essential Companion (Streather)
Porsche 996 'Supreme Porsche' – The Essential Companion (Streather)
Porsche Racing Cars – 1953 To 1975 (Long)
Porsche Racing Cars – 1976 On (Long)
Porsche – The Rally Story (Meredith)
Porsche: Three Generations Of Genius (Meredith)
RAC Rally Action! (Gardiner)
Rallye Sport Fords: The Inside Story (Moreton)
Redman, Jim – 6 Times World Motorcycle Champion: The Autobiography (Redman)
Rolls-Royce Silver Shadow/Bentley T Series Corniche & Camargue Revised & Enlarged Edition (Bobbitt)
Rolls-Royce Silver Spirit, Silver Spur & Bentley Mulsanne 2nd Edition (Bobbitt)
RX-7 – Mazda's Rotary Engine Sportscar (Updated & Revised New Edition) (Long)
Scooters & Microcars, The A-Z Of Popular (Dan)
Scooter Lifestyle (Grainger)
Singer Story: Cars, Commercial Vehicles, Bicycles & Motorcycle (Atkinson)
SM – Citroën's Maserati-engined Supercar (Long & Claverol)
Subaru Impreza: The Road Car And WRC Story (Long)
Supercar, How To Build your own (Thompson)
Taxi! The Story Of The 'London' Taxicab (Bobbitt)
Tinplate Toy Cars Of The 1950s & 1960s (Ralston)
Toyota Celica & Supra, The Book Of Toyota's Sports Coupés (Long)
Toyota MR2 Coupés & Spyders (Long)
Triumph Motorcycles & The Meriden Factory (Hancox)
Triumph Speed Twin & Thunderbird Bible (Woolridge)
Triumph Tiger Cub Bible (Estall)
Triumph Trophy Bible (Woolridge)
Triumph TR6 (Kimberley)
Unraced (Collins)
Velocette Motorcycles – MSS To Thruxton Updated & Revised (Burris)
Virgil Exner – Visioneer: The Official Biography Of Virgil M Exner Designer Extraordinaire (Grist)
Volkswagen Bus Book, The (Bobbitt)
Volkswagen Bus Or Van To Camper, How To Convert (Porter)
Volkswagens Of The World (Glen)
VW Beetle Cabriolet (Bobbitt)
VW Beetle – The Car Of The 20th Century (Copping)
VW Bus – 40 Years Of Splitties, Bays & Wedges (Copping)
VW Bus Book, The (Bobbitt)
VW Golf: Five Generations Of Fun (Copping & Cservenka)
VW – The Air-cooled Era (Copping)
VW T5 Camper Conversion Manual (Porter)
VW Campers (Copping)
Works Minis, The Last (Purves & Brenchley)
Works Rally Mechanic (Moylan)

www.veloce.co.uk

First published in August 2008 by Veloce Publishing Limited, 33 Trinity Street, Dorchester DT1 1TT, England. Fax 01305 268864/e-mail info@veloce.co.uk/web www.veloce.co.uk or www.velocebooks.com.
ISBN: 978-1-84584-174-4/UPC: 6-36847-04174-8

Readers with ideas for automotive books, or books on other transport or related hobby subjects, are invited to write to the editorial director of Veloce Publishing at the above address.
British Library Cataloguing in Publication Data - A catalogue record for this book is available from the British Library. Typesetting, design and page make-up all by Veloce Publishing Ltd on Apple Mac.Printed in India by Replika Press.

THE STORY OF THE **UP-N-SMOKE** ENGINE PROJECT • DANIEL PEIRCE

The Fine Art of the

MOTORCYCLE ENGINE

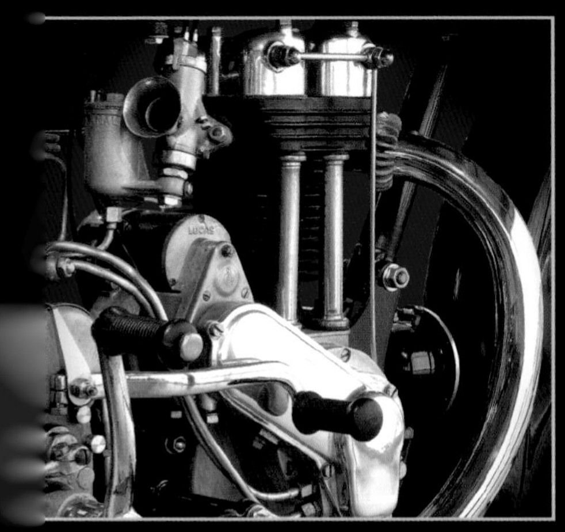

VELOCE PUBLISHING

THE UNS ENGINE PROJECT: FOREWORD

"Pornography for Gearheads" – *Cycle World* magazine

When Daniel asked me to write the foreword to his first book, I first felt honored. Then I thought, "Well ... what can I say about him that he hasn't already said about himself, and in far more glowing, eloquent terms than I could ever manage?" Self-proclaimed 'Lord of Light, Master of the Pixel?' Always explaining the finer points of his photos, as if we laymen would overlook them? 'Daniel' rather than just 'Dan?' Good grief, what's next, 'Mr. Peirce?' 'The Daniel?'

Then I thought, "Yeah, but look at the pictures!" And we might also consider his sterling reputation in the motorcycle aftermarket industry, among motorcycling periodical editors, and to 'gearheads' everywhere as creator and purveyor of fine photographic images of motorcycles. Quite simply, he is one of the best motorcycle photographers in the business. There's an old saying in Texas: "If you can do it, it ain't braggin'!" Okay then, 'Lord of Light, Master of the Pixel,' and 'Daniel' it is!

Daniel and I became acquainted in the North Texas motorcycling community and weekly Peckerhead Motorcycle Racing Team 'meetings.' Soon after, we collaborated on a regular feature called 'Texas Road Runners' in the bimonthly *Ride Texas* magazine. His images and my words seemed a good fit, we liked each other, and had good fun with it. We also learned much more about each other, and about what publishers need and readers like.

Around the time that series ended I began working for the company where Daniel is Photographic Image Manager. Soon we were once again working together, this time in the creative side of producing motorcycle aftermarket accessory catalogs, magazine ads and articles, and related activities, like trade show strategies and exhibits. Our friendship grew, both at work where we kept each other amused and fanned each other's creative fires, and in our personal lives. We were lucky that our worlds included motorcycles, and we actually got paid for playing with them, something we had usually done for free. Plus, I got great professional studio photographs of my motorcycles ... life was sweet!

When Daniel first began talking about his motorcycle engine poster project, most of his friends thought it a great idea. He would sell a few posters and we would get some for our own garage walls, or maybe even have our own motorcycles selected as subjects. My advice was to include a picture of the entire motorcycle on each poster as well, so the viewer would see the engine in context. He explained that that would dilute his concept of 'engines as art.' Besides, there was already plenty of technical and historical information, and imagery, of classic motorcycles. I still thought he was wrong; but he was right.

That became clear to me when I saw the first finished prints. They were beautiful, as expected, but not just another good motorcycle picture, not just another briefly interesting cover for blank wall space where we kept our motorcycles, not just a poster you would buy at a rally and never quite get around to hanging. There was something ... more. The real proof of concept came, I suppose, when, in addition to the 'gearhead's' enthusiastic interest, even non-motorcyclists would stop, point, smile, and talk about the prints. Some also made purchases, as a gift for the lucky motorcyclists in their lives.

So, maybe he really was onto something. I was certain of it when I saw the lovely prints and ceramic tiles produced with the new silver Metallic paper, especially since he chose my beloved Norton 'Special' to be the first! But when he told me of this book, I once again thought he had taken the wrong approach. How could some sixty-odd pictures, even very good ones, of nothing but motorcycle engines, with only the scantiest historical or technical information, hold anyone's interest long enough to look closely at as a book? Or be engaging enough for them want to read the thing, much less actually buy it?

I continued to be troubled by this until I started reading the draft pages for each engine. It was pleasantly surprising to see they included not only the cool engine pictures, but a story line with colorful glimpses into his tentative quest for 'fame,' his professional and personal lives, his unique philosophy, his amusing-to-hilarious and often self-deprecating sense of humor, his warmth, his good-nature and good-will, his capacity for, and love of life. Woven through all this are interesting bits about the motorcycle subjects, their owners, and Daniel's struggle with and dedication to this project.

There is really quite a bit on each page, and this will likely be the first and only 'coffee table book' I ever actually read ... I'm eagerly awaiting the final drafts! I know you'll love the pictures, and I hope you read and enjoy the story of this book as well.

Being an optimist myself, as is our Daniel, I commented that I understood I wouldn't be paid for contributing this foreword, but would very much appreciate an inscribed complimentary copy for my collection of motorcycling books. He frowned and replied, "An inscription is no problem, and I can probably give you a small discount on the book."

Did I mention that the 'Lord of Light, Master of the Pixel' is also extremely, umm ... 'frugal?'

Dave Howe, Head Peckerhead

Dave Howe and his prized 850 Commando known as 'Nort.'

A lot of Peckerheads, but hardly all of them. This photo was taken on the occasion of Nick Burke's memorial ride.

INTRODUCTION

OR,

WHO AM I, AND WHY AM I HERE?

As a photographer I always wanted to photograph a series of tasteful female nude figure studies. I wanted to make pictures that would study the shapes and lines of the female form; and the naughty bits. However, being married to a jealous woman for 27 years, I never got around to it. As a photographer and a motorcyclist I discovered that motorcycle engines have their own seductive shapes, lines and naughty bits. So, I began the Up-N-Smoke Engine Project as a creative outlet. It was something to do in my spare time just because I liked motorcycle engines. Same as a writer might write limericks just to amuse himself.

As all too often happens, if I'm interested in something it soon becomes an obsession. Which can be convenient because otherwise I'd be pretty lazy. The Up-N-Smoke Engine Obsession took me on a two year journey that has culminated in the photography book you're holding now. This is first-and-foremost a photography book, studying the light, shadows and graphic nature of motorcycle engines. It is also the story of what I went through to produce this collection of images. Contrary to what many folks might expect, it's not filled with historical information about the engines. There is some information about the engines on these pages – but not a lot. I'm just a photographer, and I'll leave the empirical data to the motorcycle historians.

The engines for this book were selected on three criteria. First, the engine's graphic appeal. The engine had to intrigue me in some way by its shape, balance, complexity or simplicity. Secondly, I wanted some of the more historically notable engines in the collection to represent the changing technologies and design trends through time. And the last criterion was whether I could actually find these engines to put in front of my camera. It's because of this last criterion that some of your favorite engines and mine are not represented in this collection. You'll also note that very few of these engines are factory correct. As an art project I like the engines to look like they have their own personal history. It makes you wonder what they've been through, and provides a reason to look deeper. So, I present to you a visual study of the motorcycle engine – not every motorcycle engine, but quite a few of them ... so many engines, so little time.

For good or bad, I've written this book in my own voice. It's written in the same manner in which I speak, and Texans are known for speaking an idiosyncratic form of English. My Microsoft Word grammar check tells me that I've likely angered the god of sentence fragments. Pretend we've just sat down to enjoy a beer and you made the mistake of asking me to tell you the story of the Up-N-Smoke Engine Project. In fact, reading this book while drinking a beer or two is the proper way to enjoy it.

Since this is a personal story, I think a brief bio is in order. I was born Daniel William Peirce in 1959 in Washington DC. My father was a career soldier in the US Air Force, and we were a typical military family – which means we moved a lot. Since the age of four I grew up in Universal City, which is just outside Randolph Air Force Base, which is just outside of San Antonio, Texas. Universal City was a small town of 10,000, made up of a mixture of military and rural folks. I graduated in 1977 from Samuel Clemens High School where my guidance councilor advised me to become a plumber. It was a good suggestion and I'd probably be one today if I hadn't wanted to take a stab at photography first. I didn't go to college, but, in 1980, I was smart enough to marry a woman who did. We had two children, Elizabeth, now 24 years old, and Catherine, 19 years old at the time of writing. We moved from Universal City to the Dallas/Ft. Worth area in 1985, then to Hawaii in 1992, and back to D/FW in 1995. In Hawaii, I managed a photo lab for a Japanese company. (The Hawaii adventure will be a whole other book some day.) During my photographic career I've been a pet photographer, police forensic photographer, sports photographer, wedding and portrait photographer, photo lab manager, and photography teacher. In between all that I was always a commercial photographer trying to pick up advertising and magazine assignments. My wife Pam and I live today in beautiful Grapevine, Texas, and, at the time of writing, I'm in charge of catalog and advertising photography for Tucker Rocky Distributing, where I've been for 12 years.

I dabbled in motorcycling in my teens, but after I got married Pam warned me strongly against riding motorcycles. By the time I was 38 years old the girls were old enough, and I had enough life insurance, that Pam relented to my desire to have a motorbike again. I rebuilt a free 1972 Honda CL350 basket-case and started riding again. A few years later I bought a 1978 Honda CB750 that became my daily rider. I think I enjoy working on motorcycles maybe a little more than I like riding them.

Glossary

To help with some of the more unfamiliar terms used in this book, I've provided these definitions:

Landshark – The name of my 1978 Honda CB750, derived from the Flying Tiger shark grin painted on the tank.

NTNOA – North Texas Norton Owners Association. An association of people who appreciate Nortons, as well as all British and European motorcycles in general. Many NTNOA members are also Peckerheads. www.ntnoa.org

Peckerhead – 1. A Southern US term for a woodpecker. 2. A derogatory term for a person with questionable social skills. 3. A member of Peckerhead Motorcycle Racing, a group of motorcyclists who meet every Friday evening at Up-N-Smoke Barbecue House. The organization has no dues, no rules, no bylaws, no executive officers, no membership list, and it never races motorcycles. To qualify to be a Peckerhead, a member must show up at Up-N-Smoke on Friday every now and then, be able to talk motorcycles to some degree, demonstrate the ability to drink large amounts of beer (optional), and have questionable social skills. All motorcycle brands are accepted; from American Harley-Davidson, to German BMW, to Japanese Yamaha, to Chinese Chang Jiang. The group started meeting almost twenty years ago every Friday evening at Dave Howe's motorcycle hobby-shop, behind his house. Years later, members would meet first at Up-N-Smoke to eat, drink and tell lies, then they'd pack things up and ride over to Dave's place to finish the evening. In 2006, Dave retired and moved the Peckerhead International Headquarters to his new home near Amarillo, Texas, where a new chapter of Peckerheads has been formed. The Peckerheads left behind in Keller still meet every Friday at Up-N-Smoke.
<peckerheadracing.netfirms.com>

Up-N-Smoke (UNS) Barbecue House – A motorcycle and racecar themed restaurant and bar located in Keller, Texas. Phil Dansby, the owner, doesn't mind the Peckerheads meeting there because he too is a Peckerhead. Great food and 29 varieties of beer on tap.
<www.upnsmokebbq.com>

Special thanks to the Barber Vintage Motorsports Museum for its cooperation in allowing me to photograph motorcycle engines in its collection. Please visit the museum in Leeds, Alabama, USA.
<www.barbermuseum.org>

FEATURED ENGINES

Please note: segments are listed by the artist's titles for the photographs and do not necessarily include the make or model of motorcycle.

Dedication

This book is dedicated to all the women I must please. To Pam, Elizabeth, Catherine, Nancy and Michaela ... and, of course, to Peckerheads everywhere.

Daniel Peirce

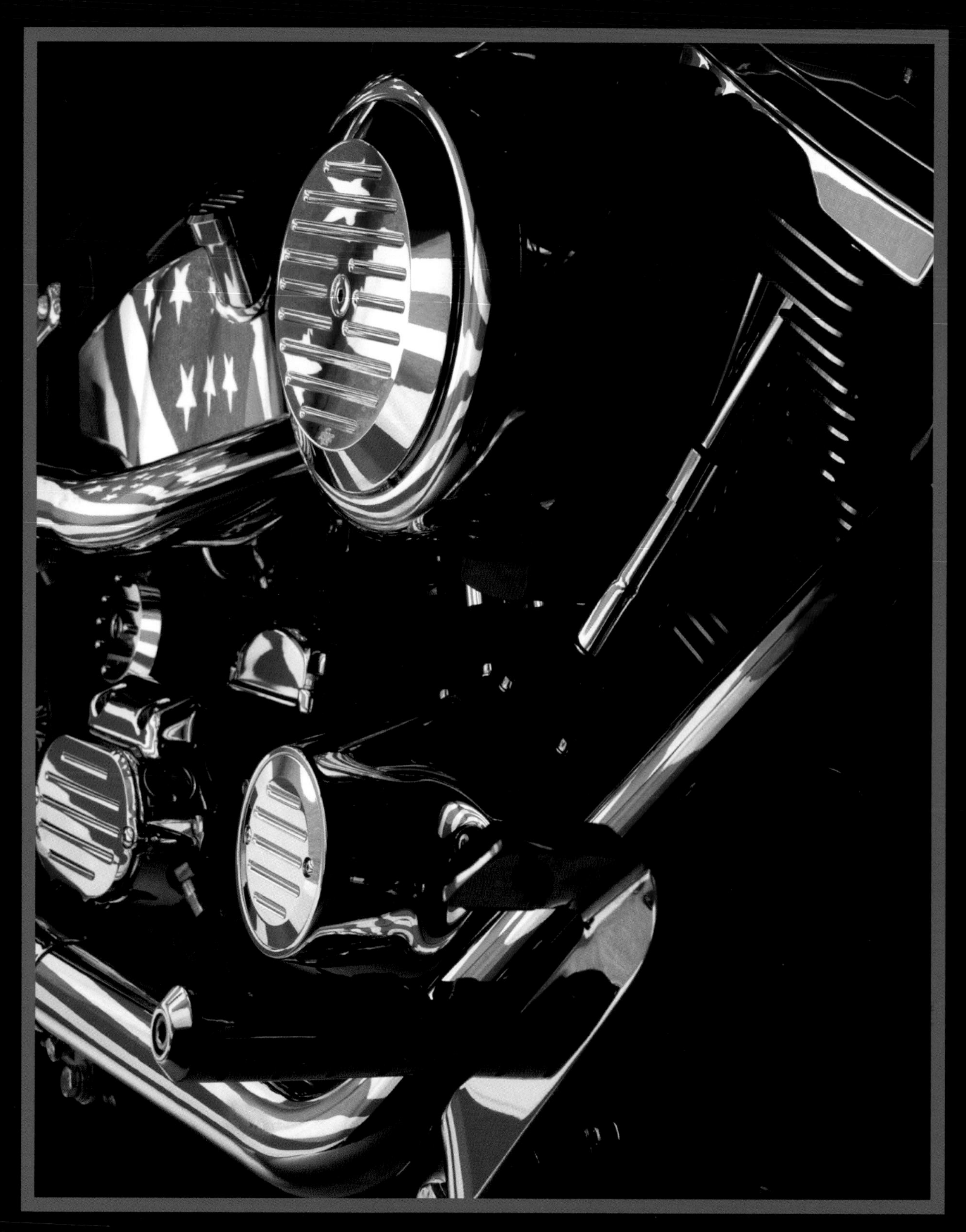

(Motorcycle courtesy Jan Milburn)

AMERICAN BEAUTY

1995. The American Beauty (subtitled: The American Evolution) wasn't the beginning of the UNS Engine Project. More accurately it is the project's ancient ancestor. It is included in the collection because it was the first engine I photographed solely for its graphic appeal.

It was intended to be the cover shot for the 1996 Nempco catalog. Nempco was an American V-twin aftermarket parts distributor that eventually became the present day Biker's Choice. On a hot summer day, in a warehouse in Texas, my job was to spend a day photographing a nicely customized Harley-Davidson to come up with something eye-catching for the cover.

i photographed the bike from every angle I could think of. I used every technique I could think of. I used every lens and every light I had. By the end of the day I knew I had some pretty good stuff, and I was confident that a good cover would come from one of the images. As I started to pack up I had a sudden inspiration. I draped an American flag over one of my broad lights so it reflected in all the chrome bits. I shot an entire series of pictures using the flag, but the chromed-out Evolution engine gave the best effect. "I'm a genius," I thought to myself ... once again.

When I presented transparencies to Tom, the art director, I waited for the moment he would see the flag-reflected engine picture. I waited for the "ooouuhhhh" that he would undoubtedly utter when he realized its creative perfection. But there was no "oooouuhhh" when he looked at it. Tom examined it for a while and simply slid the transparency to the discard pile. Somewhat shocked, I commented, "Hey, I thought that would make a dandy cover shot." Tom, still looking through film on the light box, matter-of-factly said, "Yeah, but it's a little corny." 'Corny?!' I was more than a little disappointed. I would have accepted, "Pandering to the stereotypical relationship of patriotism and American-made V-twin motorcycles." But not 'corny.'

The picture was never used. Art directors always win over photographers. I rescued the image from the discard pile and filed it for my portfolio. The Up-N-Smoke Engine Project was still ten years away ...

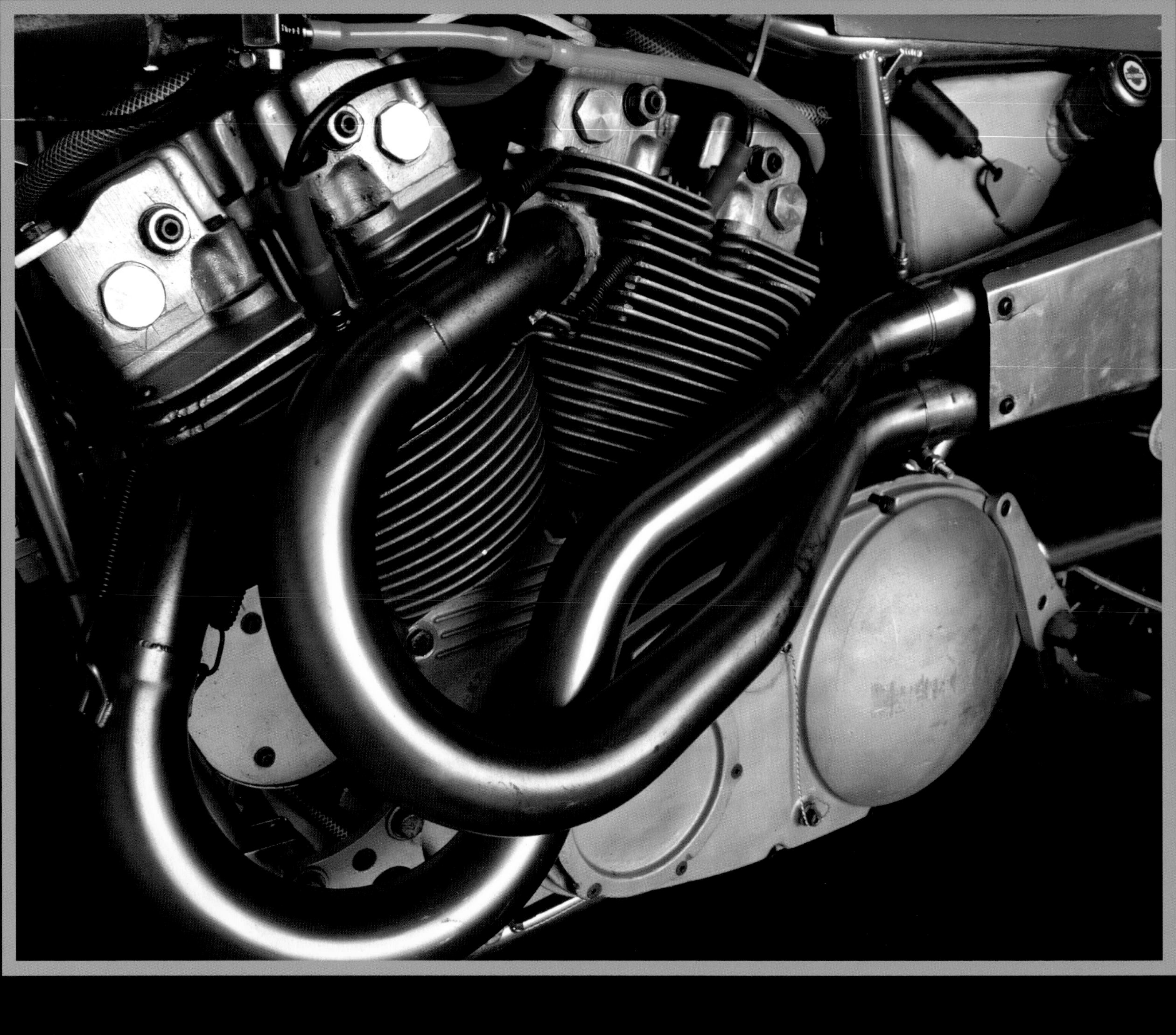

(Motorcycle courtesy Bill Milburn)

XR-750

Photographed a few months after the American Beauty, the XR-750 still wasn't the beginning of the engine project. This XR-750 engine was photographed for a feature article about the race bike for the June 1997 issue of *Hot Bike* magazine. Some of the engines I photograph I know almost nothing about, but because of the magazine article, I know quite a bit about this one.

Collector (and notorious wild man) Bill Milburn owns the restored 1975 XR-750 factory dirt tracker. It was originally ridden in the mid-70s by none other than dirt track racing legend Mert Lawwill on the factory racing team. In the 1980s the bike was handed over to an up-and-coming star named Chris Carr. Even when Carr was riding the bike, Lawwill continued to do the tuning. Hard to see from the outside, but Mert added many of his own performance parts to keep the bike competitive. The ground-up rebuilding was undertaken by Gerard Jessup. Jessup was careful to keep the tracker in top performance condition. He had to, because he was still sliding the bike around the track while racing with the Vintage Dirt Track Racers Association.

I refer to this picture as the 'well-used' XR-750. And that's part of its attraction. The engine looks like it just came off of the track, and it's ready to go right back on. The problem with this engine is that it looks great on both sides. Which side to choose for the picture? For me, it's those pipes. The headers on this all-business race bike are undeniably sexy. Marry the flowing lines of those pipes to an engine with veteran character and you have a picture that demands to be studied.

Motorbikes from A to Z, Lake O' The Pines 2001

MOTORBIKES FROM A TO Z, LAKE O' THE PINES 2001

'Motorbikes from A to Z, Lake O' The Pines 2001' is not part of the UNS Engine Project collection. However, it is the seed from which the project grew.

Lake O' The Pines, near Jefferson, Texas, is the site for the North Texas Norton Owners Association's annual rally. It's an October weekend of riding, telling lies and relaxing in the shade with a cold brew. Tall pines shade the campgrounds and yield an idyllic view of the lake. The Texas summer heat has broken and the trees are just beginning their fall colors. An exceptional location for sipping beer and shining chrome.

2001 was my second trip to the Lake O' The Pines rally. I made special arrangements to do an article on the event for *Ride Texas* magazine while I was there. I would simply squeeze it in between my riding time and my beer drinking time. The highlight of the event is the Saturday vintage bike show. The show features primarily British and European motorbikes, but this year they added two small categories for American and Japanese models. For some reason, this show brings out some of the best vintage motorcycles from all over Texas.

The lake is the background for the bike show, and bikes are lined up by categories and snake along the shoreline. My camera and I began wandering the show looking for pictures to illustrate the article I was producing. With all these remarkable motorbikes in one place I began to notice the individual characteristics that made each one unique. Soon I was on my knees focusing my camera on all the interesting motorcycle bits that defined each marque. It turned out to be one of the most enjoyable photography sessions of my career.

The next year I selected ten of the best images from that session and framed them as a two print collection. I donated the framed set to the club to be given away as a door prize. To my surprise I received orders for several more sets from other rally-goers.

While assembling the orders it occurred to me that it was mainly the engines that I was attracted to. Hmmm ... it started me thinking. But I couldn't think about it too long; this was during a period where I was busy photographing for magazine articles; no time for art. The birth of the project was still a few years, and a couple of significant pictures, away.

Early morning mist rises off of the water at the Lake O' The Pines Rally.

(Motorcycle courtesy of Phil Dansby)

EYE CANDY

Phil Dansby owns the Up-N-Smoke BBQ Restaurant and Motorsports Bar in Keller, Texas. The project, for various reasons, takes its name from his establishment. Phil is also an accomplished and noted European motorcycle restorer and collector ... and a Peckerhead, but that's another story.

In 2000, Phil called me to ask if I would photograph his pair of 1973 Norton Commandos for an article in *Cycle World*. At that time I wasn't doing much photography outside of my day job, but a chance to have my work appear in *Cycle World* was attractive (not to mention the fair amount of free barbecue Phil threw into the deal).

David Edwards, Editor in Chief at *Cycle World* liked the studio photography of the bikes, and it made a first-rate article. That was when I started thinking seriously about producing motorcycle photography for magazines as a sideline. Less than a year later, I established Trick Photography as my part-time business.

When Phil finished his newest bike project, he called me again for another round of photographs for *Cycle World*. I'm a sucker for free barbecue. Phil's new creation was appropriately dubbed 'Eye Candy.' He had constructed a very sexy replica of a 1970s Ducati Desmo F3 factory road race bike. Except he pulled a 450cc engine from a 1974 Ducati RT to power it. Phil doesn't stop building until he reaches perfection, and the Eye Candy bike is his masterpiece. He even added the crowning touch of a Gear Gazer over the bevel drive that indicates proper oil temperature for racing.

The Eye Candy engine photo was just one of the series of studio shots that were submitted to the magazine. The article only used one shot of the full bike, but my favorite picture was of the engine. It was a desmodromic metal sculpture. Undeniably, this was art. Too nice a picture to just tuck it away in my files. This would make a great poster, I thought to myself. But I didn't know anything about producing, publishing and distributing posters (and publishing also sounded expensive). I reluctantly tucked the Eye Candy image into my files.

Phil's masterpiece, Eye Candy.

XR750 UNS

In 1979, I was 20 years old when I first saw the science fiction movie *Alien*. The movie had an astounding impact on me. And it wasn't just because I was stoned at the time. *Alien* was art directed by H R Giger, who not only designed and created the alien monster, but all the sets and spacecraft as well. His art direction combined organic and mechanical forms in the sets, props and costumes. Giger's airbrush art already had a reputation for merging mechanical forms with surrealistic erotic and organic shapes. His style was dubbed 'biomechanical,' and he brought this style to bear in the *Alien* production. I didn't realize it at the time, but his art had a lasting influence on my photography.

cs ~ ɛɔ

In 2005, Phil invited me to a party at his home the night before his restaurant hosted its fall car and bike show. The party was for Phil's friends who came from out of town to display their motorbikes in the show. When I got to the party I found everyone in the garage standing around a very special XR750 street tracker.

There it was. Not only was this the perfect XR750 engine, it also had Giger's sexy biomechanical lines. The pipes seemed to grow right out of the cylinders. Everything about the engine had a biological symmetry; it may have been built to perform, but its design was pure art.

The bike belonged to Randy Porter, and is a masterpiece from famed motorcycle builder Ron Wood. Ron initially built this XR750 for Ricky Graham to race in the early 1980s. In 1989, Dave Reese bought the bike and had Wood rebuild it for street riding. Ron fell in love with the project and made what has been called, 'the most finely detailed motorcycle ever built.' Ron personally dished the head of every bolt on a lathe before they were re-plated.

I took this picture the next day in the parking lot at the Up-N-Smoke Car and Bike Show. Now I had four very nice pictures of motorcycle engines. I was beginning to see a trend here. I knew the history of motorcycles included many stylish engine designs, and I figured that there had to be a lot of gearheads in the world who appreciated the look of a naked engine as much as I did. Thus, I quietly began the engine project.

The first print ever made of this beautiful XR750 engine hangs in the Up-N-Smoke BBQ restaurant and Motorsports Bar, Keller, Texas, USA.

(Motorcycle engine courtesy Phil Dansby)

MATCHLESS

There's an old saying in Texas: "Shoot first and ask questions later." Well, that sums up how my new project started. All I knew was that I was going to produce a body of work featuring artfully crafted photographs of attractive motorcycle engines. What I was going to do with this body of work was a question I still didn't have an answer for. I had vague ideas about selling prints, and maybe even a book. But you can't do anything with something that doesn't exist. I would shoot first and ask questions later.

Now I needed interesting motorcycle engines to photograph. Once more, Phil Dansby and his Up-N-Smoke restaurant enter the scene. The walls of the barroom are decorated with several engines cut in half and mounted on boards. BMW, Royal Enfield, BSA, Norton, Matchless and others grace the walls of the motorcycle-friendly Up-N-Smoke. This seemed like a good place to start. The only problem was that the engines were too heavy, and too difficult to take off of the wall to take to my studio. I would have to photograph them while they were still bolted to the barroom walls.

I made arrangements with Phil so I could come in early on a Sunday morning before the restaurant opened. My wife Pam went with me. She's great at holding reflectors ... and she likes good barbecue and cold beer as much as I do. After an hour of photographing while balanced on bar stools I had gleaned what I could from the collection.

When an art project like mine is being produced you have to decide on a general consistent look. The consistent look of my project was still evolving. I liked the strong lines and graphic appeal of this close-up of the Matchless. I still like it. But, it's the only real close-up I have in the collection. The subtle colors you'll find in the metal in this picture come from the lighted neon signs that share the walls of the bar with the mounted engines. The red of the Matchless logo provided the color for the border lines.

I was still trying to come up with a catchy name for the project.

The walls of the Up-N-Smoke barroom are decorated with a variety of motorcycle treats, including several mounted half engines.

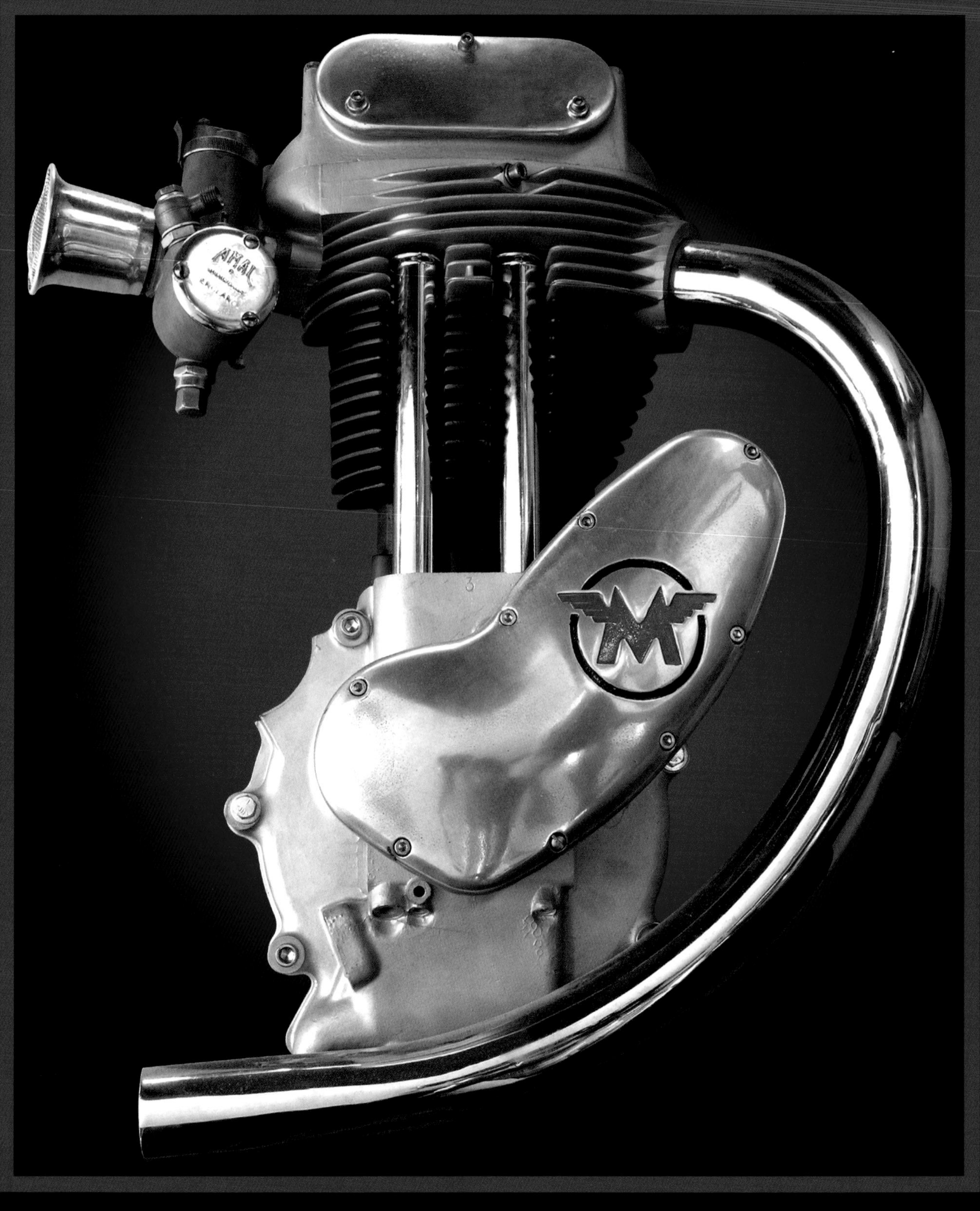

(Motorcycle engine courtesy Phil Dansby)

MATCHLESS UNS

Phil liked the close-up of the Matchless I made, but he thought a picture of the entire engine would be more appealing. He requested that I also make a print of the full engine. I'm easy. The result was the Matchless UNS. I have to admit that it's a handsome motor. I subtitled the Matchless UNS, Phil's Full Monty.

The Matchless half engine wall hanging was conceived and assembled by Phil Dansby, but legendary racer/builder Ed Mabry was responsible for cutting the Matchless 500cc single in half. I've known a lot of great motorcycle people through Peckerhead Motorcycle Racing, but Ed is special. His life is racing, and his success with motorcycle land speed records is well documented. Ed has the humility and down-to-earth character of a true Texan, and the smarts and ambition of ... well, a true Texan.

This picture was also responsible for helping me find a name for the project. I had already titled the close-up picture as Matchless. Now what was I going to name another Matchless picture? Well, the picture was taken at Up-N-Smoke, and it was made at Phil's request who owns Up-N-Smoke; so I decided to recognize that by adding the initials UNS to the title.

Then I started thinking that UNS could be confusing to people reading the title. They'd be wondering what UNS stood for ... cool; I like a little mystery in my work. The name for the entire project should have a little mystery as well. I chose the UNS Engine Project as the project's name. Later, I started alternating UNS with the longer Up-N-Smoke.

I realized that Up-N-Smoke was an appropriate name for several reasons. To begin with, it was the restaurant where the idea for the project was born. The very first picture from the project was hanging in Up-N-Smoke, although that print had no title other than my name. I later changed the title of the perfect XR750 to the XR750 UNS. And lastly, Up-N-Smoke can refer to the fact that most of the vintage engines in the project are now merely ghosts in the ethereal smoke of memory and history.

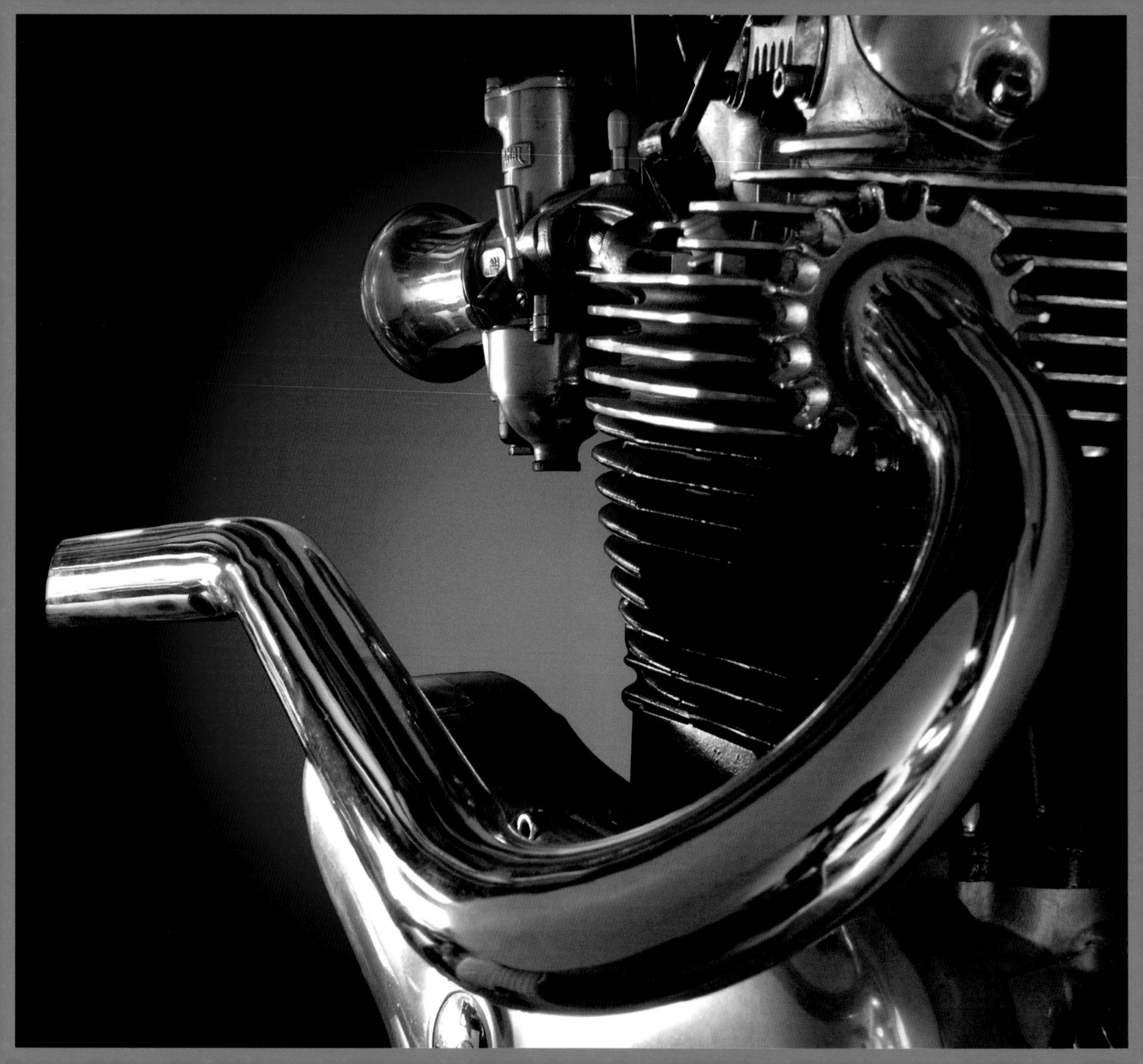

NORTON UNS #1

I guess I figured that I'd be making a lot of pictures of Norton engines. Either that or I just couldn't come up with a better title. But there never was a #2. I could change the title, but once a picture is named it takes on a personality. It's a lot like naming a child.

This is another of the cut and mounted engines decorating Up-N-Smoke. It's a Commando Combat head mounted to a Dominator engine case. The high-pipe Commando wall hanging was constructed in a collaborative effort between Phil Dansby and Dave Howe. Once again, you can detect the colors of the neon bar lights in the reflections in the metal.

cз ~ ɛɔ

Seemingly unrelated, a co-worker, Will Pattison, returned from a trip to Alabama, raving about a newly-built motorcycle museum with an incredible collection of bikes. Located just outside of Birmingham, the Barber Vintage Motorsports Museum had just moved into a new, custom-built facility. The pictures Will showed me validated his ravings. The museum and its collection were impressive, indeed. Will continued to rave, "You gotta' go, Dan. You'd love it ..." Yeah, well, maybe someday.

The Up-N-Smoke BBQ House in Keller, Texas, USA.

(Owner unknown. Photo taken randomly at the Fall 2005 Up-N-Smoke Car and Bike Show. I need to start remembering to find out who owns the bikes I was photographing.)

SPORTSTER CHOPPER

This is not a Shovelhead engine, and no real motorcycle person would ever mistake it for a Shovelhead; yet I erroneously titled it as a Shovelhead. It carried this title for months before someone finally pointed out my mistake. It's a Harley-Davidson Sportster engine in a chopper frame.

Instead of just re-titling the picture, I pulled it out of the collection. The image simply lacked the drama I wanted an engine to display. I've included it in this book because it's part of the story of the project. But you won't find it in the collection. It was simply a victim of a photographic style that I was still trying to define.

☙ ~ ❧

I was starting to run out of engines to photograph. I decided that, if I was serious about producing the project, I needed a pool of engines to take pictures of. I committed to spend the time and money to travel to the Barber Vintage Motorsports Museum in Leeds, Alabama.

Money was always a big factor in the project – principally because I didn't have any. At this point I was still spending more money than I was taking in. And that's because I wasn't taking any money in. But that's the difference between a project and a business. I already knew I was a terrible businessman. I naively believed that, if I concentrated on making good pictures of cool engines, the money would take care of itself. Or not; I make everything up as I go along.

I sent an email to Jeff Ray, the executive director of the museum, and described my intentions. We exchanged emails for a few days, but we only established that I was coming, and that I wanted to photograph the motorcycle collection. He didn't tell me not to come. I marked a date in April, 2005, to visit the museum and began making plans for the trip.

(Motorcycle courtesy of Barber Vintage Motorsports Museum)

SCOUT

I was determined to travel the 700 miles to Leeds, Alabama, to photograph motorcycle engines at the Barber Vintage Motorsports Museum. I had a twelve-year-old Mazda dentmobile with 145,000 miles on it, but I was feeling lucky and I was going to trust it to survive the journey. If I ate light and found cheap motels I could probably complete the trip in five days and 300 US dollars. Not a prohibitive sum, but still a big investment in something I wasn't sure would ever have a pay-off. My life has always been filled with leaps of faith. Maybe that was why I didn't have any money.

Enter Jim Bagnard, my riding/drinking/wrenching buddy and occasional business associate. Jim had been fairly successful as a part-time motorcycle industry entrepreneur, but his day job was as a captain for American Airlines. When he found out that I was planning a trip to the Barber Museum he decided to go with me, just to visit the museum. Jim appreciates vintage motorcycles and has a special passion for Indians.

Airline pilots receive a certain amount of flight benefits for traveling companions. He was able to arrange discounted airfare for us to fly into Birmingham; cool. The trip would still probably cost me the same amount of money, but it would save me a long drive in a dodgy automobile.

When we arrived at the museum we paid our admission and I asked the ticket lady if I could speak to Jeff Ray, the museum's executive director. I was basically a nobody, just a photographer who had emailed some weeks before asking to come by and take some pictures, so I wasn't surprised when he couldn't exactly recall who I was or why I was there.

Jeff was a tall man with a pleasant face and the politeness that southerners are famous for. But underneath I could tell that he was all business. I explained about the project and told him that I planned to eventually sell the images as posters so I wanted his permission before I started shooting. He considered this for a moment and said there would be two rules: I was not to shoot any pictures of entire motorcycles, only details; and I couldn't shoot any pictures at all of the Britten that was on display. Perfect ... I went to work.

~

The Scout had the biomechanical lines I always look for. Indian always had a flair for style with its motors.

(Motorcycle courtesy Barber Vintage Motorsports Museum)

SANDCAST CB750

The Barber Vintage Motorsports Museum is designed to impress. The five story modern building sits on a campus complete with a first-class road track, and dotted with large distinctive sculptures. Inside, there are three floors of display area that are joined by broad spiral walkways. The building's center is dominated by a large glass elevator that gives access to a basement where bikes are stored and worked on. Despite a general concrete décor, large window areas give the space a rich and open feeling. I was impressed ... and I'm hard to impress.

Jim would occasionally help me shoot by holding a reflector, but he was primarily there as a vintage motorcycle fan. The Indian was Jim's favorite marque of interest. He owned a 1948 Indian Chief, so, after visiting the area where the Indians were displayed we moved on to my area of interest – Honda.

I honestly don't have strong loyalties to any particular brand of motorcycle. But for some reason I've always owned Hondas (except for a Benelli 125 my brother gave me when I was 14). Always lacking available funds, I have yet to buy a brand new motorbike. Instead, I buy them old and cheap, and rebuild them. Luckily, I like rebuilding old motorcycles. My personal daily rider was a 1978 Honda CB750 K8 SOHC, that eventually became known as the Landshark. It was originally called the Whore. When I first got it I worked on it so much that my wife, Pam, would come out to the garage and say, "I knew you'd be out here with your whore!"

When I came across the museum's 1969 CB750 K0, I knew it had to be in the collection. The K0 bike used sandcast engines. The rarity of those first sandcast engines makes it a distinctive and sought after collector's item.

I had entered the museum with a mental list of 'must have' engines: Brough Superior, Vincent Black Shadow, and the like, but I never considered a Japanese engine until I saw the sandcast CB750. Even though I appreciated Hondas, would anybody else? My decision has been validated since then by the fact that the sandcast CB750 print is one of the best-selling prints in the series.

The infamous Landshark.

(Motorcycle courtesy Barber Vintage Motorsport Museum)

SUPERIOR

High on my list of 'must-have' engines was 'the Rolls-Royce of motorcycles, the Brough Superior. The beautiful SS100 model featured the handsome and daunting JAP engine. In its day it was one of the sexiest motorbikes on the road.

ଔ ~ ଓ

By the summer of 2005, I was taking my time turning out the pictures. I was carefully making the images from the Barber Museum into pieces of art for the project. Although I have a sense of what the final look of the picture will be when I photograph an engine, I don't really know how it will look until I finish it. On average it takes me about three hours to complete the retouching and formatting for each picture. Working in my spare time on my computer at the studio, it typically took me a week or two to finish an image.

Of course I could do it in one three hour stretch, but working in small incremental steps gives me a chance to think about what I wanted to do next. The engines already have the lines; I just had to supply the drama.

The pictures from the Barber Museum were going slowly, but I wasn't in a hurry. I was still making pictures just to please myself. It occurred to me that maybe the people at the museum would like to see the results of my work. It could be a good idea to find out what they thought of them. The collection was still small, but I was already showing off what I had to anyone who would look at them. I decided to complete one more picture before I let the museum see what I had.

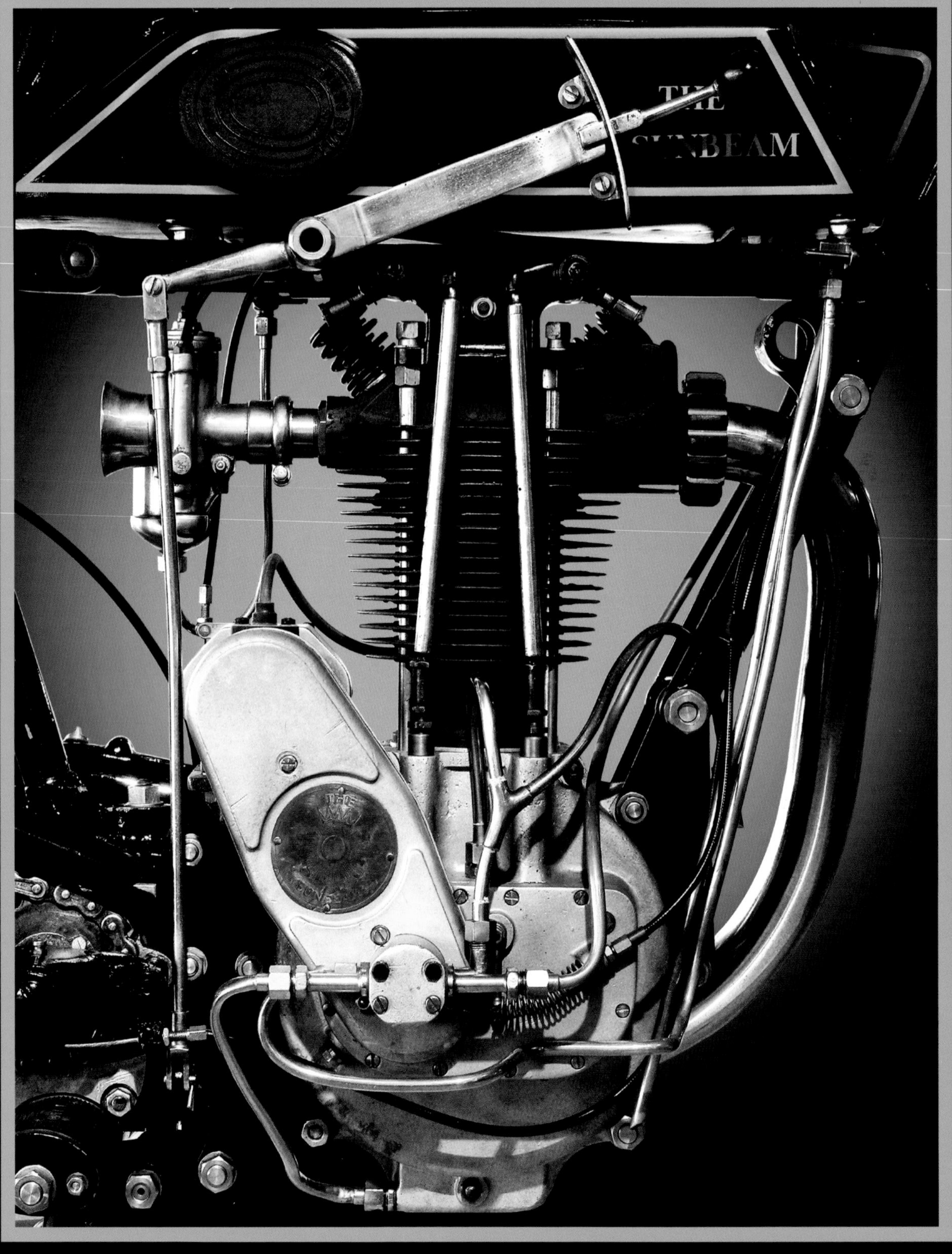

(Motorcycle courtesy Barber Vintage Motorsports Museum)

THE SUNBEAM

I like to call this one the 'Plumber's Nightmare.' The intricate external oil lines made The Sunbeam irresistible to photograph. I usually know something about the engines I shoot before I find them. Not this one. The Sunbeam was photographed just because it's graphically interesting. I also liked the fact that the emblem on the tank said, 'The Sunbeam' instead of just Sunbeam. I don't know of any other marque that does that.

With the completion of The Sunbeam picture I decided to email Jeff Ray at the Barber Museum. I sent him image samples of the Scout, Sandcast CB750, Superior, and The Sunbeam. Jeff thought they were great and suggested that they might try selling them in the museum's gift shop. Now there was a good feeling. A museum wanted my pictures. I was starting to think of myself as a real artist.

I asked Jeff if he wanted to give me the museum's logo to put on the prints. He declined giving out the museum's logo, but he did ask that I credit the museum as the source for the engine subject. That wasn't a problem with me. Giving the museum credit on the print would help give my work credibility. I also appreciated the museum working with me, and I was eager to make my work provide a benefit for them as well. I told Jeff that I would mention the museum in any publicity the project might receive in the future. Jeff suggested that a donation of a set of prints to the museum's library would be a welcome gesture. No problem there either.

However, with this success I had a problem. How was I going to produce poster prints that would be inexpensive enough to sell in a gift shop? Up until now I'd been having prints made one at a time. I'd been a commercial photographer for long enough to know that cheap prints mean quantity, and quantity always involved large lumps of cash.

There was that money thing again. My task now was to find out how to get these images printed in quantity – while finding the money to pay for it.

(Motorcycle courtesy Barber Vintage Motorsports Museum)

ROCKET

For me, BSA has the greatest variety of eye-catching engines. Not every engine it designed was a work of art, but BSA seemed to have more works of art than any other marque. The Rocket was just the first of many BSAs that would make it into the collection.

ও ~ ઙ

I searched the Internet and found three discount lithograph printers. I gave them the specifications and they gave me my options and prices. The best offer came from a printer in Iowa. To print four images as 18in x 24in color posters was going to cost me $1600.00. That would give me 500 prints of each image ... but I didn't have $1600.00.

The only person I knew who might have some investment cash was my buddy Jim Bagnard. Maybe, if I came up with a plan with a good return, he might consider investing. I drew up what I thought would be a fair and attractive plan. I made arrangements to meet Jim at Hooter's in a few days to discuss a 'business matter.'

As you may know, Hooter's is a chain of chicken wing restaurants that feature suggestively-clad young women as waitresses. It was where Jim and I traditionally held all of our business meetings. I hated to ask Jim to invest, for a couple of reasons. First, it meant that I wouldn't have complete control over the direction of my poster endeavor. Although, that might have been a good thing. Secondly, I don't like mixing friendship with money. I'd be asking him to risk his money for my idea. Risky in so many ways.

I always told my daughters as they were growing up that if you do good things, good things happen. Sort of a simplified Zen/karma kind of thing. Well, I must have done something good because something very good happened. My wife, Pam, received a financial windfall from a forgotten retirement fund. Pam asked me how much I would need to start my poster business. I figured it out to $2500.00.

Pam gave me the money. I wouldn't have to ask Jim to invest after all. Which was good because when we met at Hooter's he said he didn't have the money to invest anyway.

The touching part about all this for me was that Pam trusted me enough to invest money that would have otherwise gone to our household. Pam is a smart woman, and this was an endorsement I couldn't ignore. I ordered 2000 posters the next day.

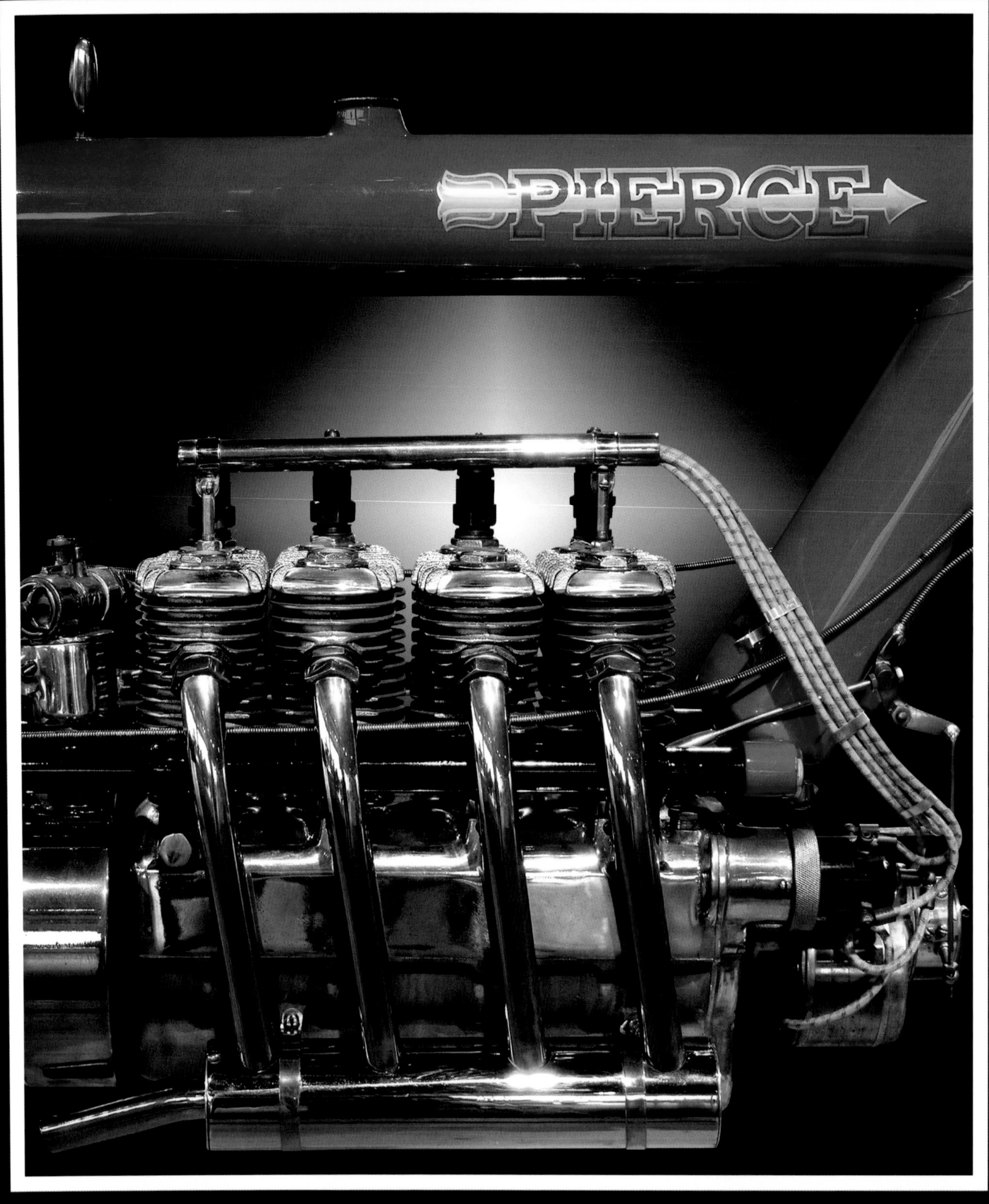

(Motorcycle courtesy Barber Vintage Motorsports Museum)

PIERCE

Now I was beginning an art project and a poster business. Unfortunately, I knew nothing about the poster business – but I had 2000 posters to sell. The Barber Museum gift shop purchased 40 of them. Only 1960 posters to go.

For some reason, I had envisioned having all the images in the project made into lithograph prints. I still had some cash from Pam's investment, so I decided to get more posters printed. I had the Rocket and Pierce images completed. How much money to get just two printed, instead of four? Answer – a lot more than I had.

I found that I could afford to get 500 prints of one image for about $500.00. I chose the Pierce, partly because it's an American-made classic that would balance out the offering, and partly because I liked the name for some reason. Except they spelled it wrong.

The museum gift shop bought ten of the Pierce lithograph. Now I owned 2450 prints that I wasn't sure how to sell. Did I mention that I'm a terrible businessman? It finally dawned on me that there was no way I could spend $500.00 to make every image into a lithograph poster. I had to find an alternative printing method that was both high quality and had a reasonable price. I needed to be able to have them printed as I needed them.

A little success can complicate your life. Instead of merely finding cool engines to photograph and making pretty pictures, now I had to find a way to sell thousands of prints, and search for a way to make even more prints. It had seemed like such a good idea at the time.

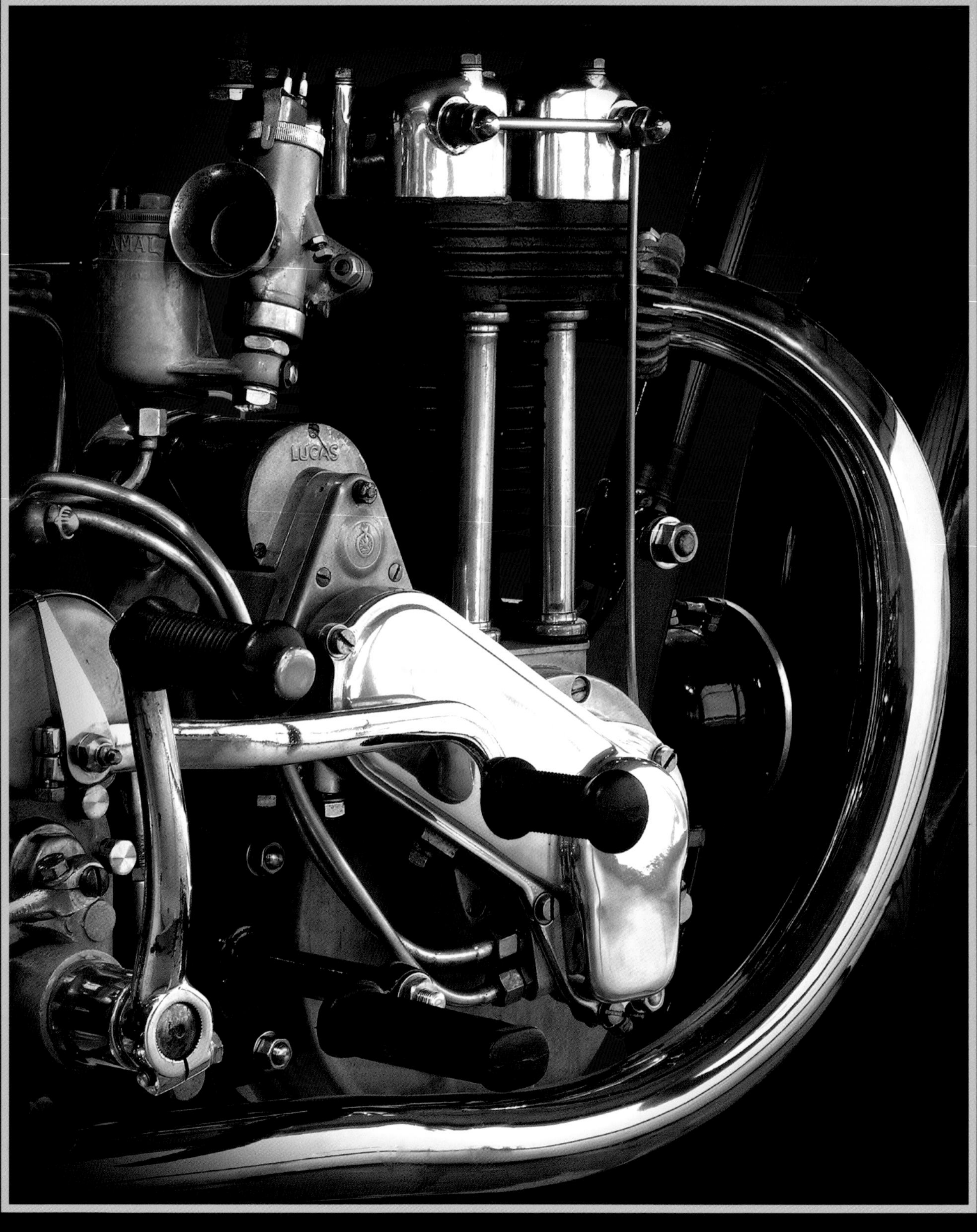

(Motorcycle courtesy Barber Vintage Motorsports Museum)

ARIEL RED HUNTER

Without a doubt, this is my favorite picture in the collection. The light, the color, and the lines of that elegant engine all work together. When it came time to title it, though, I was at a loss. I'd forgotten to write down any of the details about it while at the museum. I knew it was an Ariel, I just didn't know what model.

Searches on the internet gave me the impression that it may be a Red Hunter. But every Red Hunter I looked at was slightly different. I couldn't match the details of this engine with any other I found. Ariel had made the Red Hunter for a long time, and it went through many iterations.

I emailed the picture to Dave Howe. Dave had motorcycle contacts all over the world, and he sent the picture on to one of his guys in England. After a few days of waiting the Brit wrote back. It was indeed a Red Hunter, probably from the early 1920s. Well, I couldn't nail down the year, but the Ariel Red Hunter was a good title.

My brother-in-law finally got a job and moved out of our spare bedroom. With the money left over from Pam's investment I went about turning the bedroom into 'The Warehouse.' Shelving, tables, and the various items needed for shipping were packed into the 10ft x 12ft room. I was in business ... at least it looked like I was in business.

I knew that I would have one major problem in selling the five prints I had lithographed. They were all pretty pictures of pretty engines, but there was no broad appeal to them. The folks who would buy them would do so either for the graphic quality of a cool vintage engine, or because they were fans of the particular engine pictured. Broad distribution seemed problematic, to say the least.

(Motorcycle courtesy David Hilton)

MACH III

When I first started Trick Photography in 2001, it was primarily to photograph motorcycles for magazines. Dave Howe and I teamed up to produce a regular feature for each issue of *Ride Texas Magazine*. Known as Texas Roadrunners, the articles highlighted a different and unique motorbike in each edition. Dave did the words and I did the pictures.

One of these articles was called 'The Blue Streak Boys.' We had discovered that the Dallas/Fort Worth area was home to a group of fervent Kawasaki H1 and H2 collectors. The Kawasaki 2-stroke triples were nicknamed 'Blue Streaks' because of the blue smoke they left behind when they passed you. One of the collectors was David Hilton, a captain for American Airlines, and owner of this nicely modified Mach III (H1).

For motorcycle pictures in articles I'm all about the background. I wanted the pictures of David's bike to relate to his job as an airline pilot. David managed to get us special permission to shoot in the American Airline's hanger and on the tarmac at DFW Airport. This wasn't too long after the 9/11 attack, so getting permission for me to even set foot in a hanger was a feat in itself. Of course, this permission came with our own American Airlines public relations official to escort us wherever we went.

What made this photo shoot extra interesting was that we had to push his bike through narrow hallways and doors to get to the hanger. Security was so tight that it was the only way in or out of the building. Picture a uniformed pilot pushing his Kawasaki down the halls of an office building.

This engine picture came from that photo shoot and was used in the article. It didn't occur to me to make into a poster until David, who later became a Peckerhead, asked me, "Hey Dan, why don't you make an engine poster of my Mach III you photographed a while back?" It sounded like a good idea. I'd think about it. Then he said, "I'll buy a print of it." He had me then. The result was the "Mach III." I delivered the print to him at Hooter's.

☙ ~ ❧

It's good to have friends who are editors of magazines. I wrote a press release announcing the engine project and sent it out with a sample Scout print to four editors I knew. Three of the publications printed the press release with a picture or two. That was cool. *Motorcycle Industry* magazine just had its ad salesman call me.

David Hilton and his Mach III in a picture used for the article, 'The Blue Streak Boys.'

(Motorcycle courtesy Mike Vance)

CHIEF

Ah, October, and once again time for the NTNOA Lake O' The Pines Rally, my yearly indulgence. The rally typically draws 200 to 300 European and vintage bike enthusiasts from around Texas and surrounding states. Groups such as the British Motorcycle Owners Association from Houston, the Roadrunners from Austin, the 59ers from Dallas/Fort Worth, and another group we simply refer to as the Arkansas Boys. And of, course, my group, the Peckerheads of Peckerhead Motorcycle Racing.

I usually ride the three hours to the rally with Jim, but this year he had to miss it due to work. Instead of riding this year I put my CB750 in the back of my pick-up truck so I could bring some posters to sell. And, to save money, I decided to set up a tent in the rally campgrounds instead of getting a hotel room in nearby Jefferson.

I had camped before and swore that I'd never camp at the rally again. Many of the rally-goers camp in small group clusters. They're all friends and the beer flows freely, and after dark they just get louder and louder as the night goes on. Maybe it would be different this year ... and I needed to save money. I stopped at a hardware store in Jefferson and picked up some industrial-grade earplugs.

When I arrived at the rally most of my Peckerhead buddies were already there. I liked that; it meant that the beer would be cold. They like to come up a day early so they can ride the lake roads at their leisure. I unloaded the bike and set up my 'way-too-large,' ten-man tent. I tried to find a place to camp that was a little away from the rowdies. I took my five framed posters and set them on the ground and leaned them against my pick-up with a sign offering them for sale for $15.00.

I also brought along a table for when I planned to sell them in earnest during the awards ceremony. I don't think I mentioned that I'm also a terrible salesman, but for now I was there to enjoy myself so if somebody wanted a print, they'd have to track me down. I also took a set of prints over to the awards table to offer as door prizes.

Even though Mike Vance was a long time Peckerhead, this was his first trip to the Lake O' The Pines Rally. He didn't own a European motorbike. All he had was a 1946 Indian Chief. Okay, actually, it's a beautiful and pristine 1946 Indian Chief. Mike's Chief was fully appreciated at the rally.

I asked Mike if he minded if I photographed the engine for a new poster. I knew he'd say it was ok, but I had to ask. I photographed the Chief as it stood in the grass, while enjoying the cool breeze, in the shade of the pines, on the shore of the lake ... life was good.

Made in England
VINCENT

BLACK SHADOW

I didn't know how long I'd been asleep, but I knew it was still the middle of the night. The industrial strength earplugs had made my ears sore and woke me up. I rolled over in the sleeping bag and looked around the way-too-large tent. No sign of morning light yet. Well, surely the insanity of the campgrounds had settled down and I could get rid the earplugs. I pulled a plug out of an ear. As if I had just turned on a television with full volume, I heard, "YEEEEHAAAA!! ... OH DAMN IT! ... PICK IT UP BEFORE IT CATCHES ON FIRE!" I put the earplug back in and went back to sleep. The Norton boys were still having fun at the Lake O' The Pines.

I found out the next morning that Johnny Cool had tried to ride his Ducati through a campfire. Tried. Johnny Cool is very cool.

During Saturday's bike show I scouted for more engines to photograph; and, there it was – a Vincent Black Shadow. Now, those of you reading this who live in the UK may not regard finding a Black Shadow as a special event. But in Texas, they're scarcer than snow shovels. The Ariel Red Hunter may be my favorite engine picture, but the Black Shadow has the sexiest engine ever made. Once again, the biomechanical lines are unmistakable.

I had tried to photograph a Black Shadow at the Barber Museum, but it was close to the wall of windows and that gave me more reflections than I could retouch out. I couldn't let this one get away. I dragged out my equipment and proceeded to photograph the engine while the bike stood in the show. I'd have to find the owner afterward. Sometimes the bikes get packed up pretty quickly after the show ends.

When someone sets up a giant reflector and a camera and tripod in front of your motorcycle, you're naturally inclined to find out why. I didn't have to look for the owner; he found me. "You like my Vincent?" I heard over my shoulder. I told him that I loved it and that I was working on a photo project making pictures of motorcycle engines and trying to sell them as posters. "Would you mind if I added your Vincent to the collection?" I asked hopefully. He thought about it for a moment or two and said, "Just do me a favor and put my name on the print," and he gave me his business card. Done!

(Motorcycle courtesy Mike Thomson)

DUO-GLIDE

I started photographing motorcycles in 1988 while I was freelancing and shooting catalogs for 3D Emblem, a Harley-Davidson licensed clothing business. The first bike I ever shot in the studio was a 1958 Duo-Glide. The panhead engine, segmented pipes, and that incredible horn make a very graphic package.

On this Lake O' The Pines trip I met Mike Thomson and his wife Jessica. They were attending the rally in their motorhome and they had a trailer full of beautiful vintage motorbikes. Mike is laid-back and amiable, and welcomed my request to photograph a few of his bikes; as it turned out, it wouldn't be the last time.

Again, I was shooting a highly reflective engine under the tall pine trees of the Big Cypress campgrounds. This picture was going to take a lot of retouching. I could even see the reflections of all the folks watching me work. Retouching reflections out of chrome is always a hit-or-miss proposition for me. Luckily, this one was a hit.

The rally was winding down and the bike show awards would cap off the day. This year's special guests included two celebrities in motorcycling. T C Christenson was there showing off one of his 1970s twin Norton engine dragsters, known as the Hogslayer. David Edwards, Editor in Chief of *Cycle World* magazine was there to pay tribute to his brother who had recently passed away. David had lived in North Texas, and his brother had been a member of the North Texas Norton Owners Association.

We also had one other celebrity, 'The Texas Ceegar,' the streamliner Triumph motorcycle that set a land speed records in Bonneville in the 1950s. It was the success of this streamliner that prompted Triumph to adopt the Bonneville model name. A dedicated group from the NTNOA had just restored the Texas Ceegar after it was almost completely destroyed by a fire at the British National Motorcycle Museum. Both the Hogslayer and the Texas Ceegar were on display at the rally as their last appearance on American soil before being packed up and shipped to the rebuilt British National Motorcycle Museum.

I sat at my little table during the awards, pitifully pretending to sell posters. I sold a few, enough to pay for my trip, at least – a drop in the bucket in the grand scheme of things. I knew I'd figure it out eventually, but for now all I could do was keep shooting.

(Motorcycle courtesy Mike Thomson)

TRIUMPH 1916

This is another of Mike Thomson's bikes from the Lake O' The Pines rally. I photographed it in the same session as the Duo-Glide. Mike's pet African Gray watched me carefully while perched on the handlebars. I'm used to having spectators while I work, but they don't tend to yell, "Pretty boy!," at me. The wicker in the background is the Triumph's side car.

☙ ~ ❧

When I returned home from the rally it was time to start another project that had been brewing in my mind. My CB750 was over 25 years old and had over 70,000 miles on it. I'd hate to have that little cam chain fail on me while I was riding. Time for a rebuild.

Even though we can ride all year in Texas, winter was still a good time to work on my bike. It's a lot easier to heat a garage in the winter than to cool it in the summer. I had to pull the engine out of the frame to change the cam chain. I had disassembled a CB750 parts bike and already knew that getting the engine out of the frame was a hernia waiting to happen.

So, I figured that as long as I had the engine apart I might as well lap the valves, replace the rings and replace all the seals. I wanted to replace the primary chain set too, but I was still low on cash. I estimated that it would take me about a month to complete.

One October Friday evening in Up-N-Smoke I told the Peckerheads of my plans to rebuild my engine. I should have understood that I was making a mistake when everyone at the table just smiled and looked at each other knowingly. I looked around the table. "What? Why are you smiling?" I asked. George inquired soberly, "Have you ever rebuilt an engine before?" "Well no," I admitted, "But I have a shop manual and a lot of smart friends." They continued to smile, which I regarded as a certain measure of Peckerhead respect. Someone else and they might have just laughed ... hard.

HONDA
MADE IN JAPAN

GL 1000 WING

This is one of the few engine pictures in the collection that I don't know the motorcycle owner's name. I shot this handsome GL 1000 in the parking lot at the Up-N-Smoke 2005 Fall Car and Bike Show. Once more I was going to shoot first and find the owner later. That was the plan, but my good friend Mr Budweiser distracted me.

I mentioned before that I don't really know what the final picture will look like until I finish retouching it. This picture was a pleasant surprise. I wanted the GL 1000 engine included in the collection because it is historically significant, but I wasn't sure if it could be considered attractive. I was a little reluctant to spend time retouching it.

The finished study of the engine proved that it was indeed an attractive piece of machinery. It is stark, fluid and efficient. It's just as sexy and biomechanical as any in the project.

The Landshark, my 1978 CB750, didn't make it to the fall car and bike show. It was home on a lift table in the garage. I was still in the process of stripping off all the bits that needed to be removed before the motor could be pulled out of the frame. The rebuild had already taken a month, and I didn't even have the engine apart yet. This wasn't a good sign, but once on-task I hate to stop.

I guess I'm an optimist because somehow I thought that I could easily handle a marriage and family, a full-time job, an art project, running a poster business, and rebuilding a motorcycle engine all at the same time. Yes, I prefer to think of myself as an optimist. It sounds much better than delusional.

(Motorcycle courtesy Dave Howe)

850 COMMANDO

My buddy, and head Peckerhead, Dave Howe, had managed to get an American Flyers feature in *Cycle World* for his prized 850 Commando. He asked me if I would bring the bike into the studio and photograph it for him. No problem ... as long as I got the engine for the project. Dave had built this Norton into one very special motorbike, so I'm going to let Dave tell you its story in his own words:

"C'mon Peirce ... I had 'Nort' 30 years, and you give me 400 lousy words to tell about it? Geeze ..."

I'd wanted a Norton forever. Even nicknamed my son 'Norton.' Mine was a 1973 850 Interstate, acquired in 1975, and immediately converted to the sexier Roadster configuration.

Any new motorcycle is just a starting place for me, Nort being no exception. Modified motorcycles are my 'art' ... at least that's what I tell my wife. Early on, Nort burned in a fire and was in two floods. The tank was damaged in a repair shop, so I really had my hands full.

My vision for the bike took a while. I settled on a black, candy-red and gold paint scheme from an Excelsior Manxman, with hand lettering and lining, and styling from American flat tracker racers with solo seat and reversed pillion pad, and modern touches, like black painted forks and instruments, with polished stainless steel fasteners and other components. Upgraded brakes, suspension, rear-set controls, and modern tires made Nort safer and friendlier to ride. Improved handling and braking had become a necessity when the engine was re-done. At last ... he would turn, stop and go as well as he looked!

Ah, the engine. High compression pistons, racing top end, ported and 'flowed' head, hot camshaft, Amal Mark II carbs, improved internal breathing, adjustable cam sprocket, lightened and re-balanced crankshaft, hot coils, Boyer ignition, K&N filters, Dunstall mufflers, and more. The total re-build was accomplished by master Norton race-engine builder Ralph Delmar who later said that engine was the best he ever built.

Jack Wilson of Big 'D' Cycle and Bonneville Salt Flats fame added masterful tuning touches, then hand-laced the lovely Akront rims and stainless spokes.

It was the sweetest, smoothest, fastest, and 'happiest' Norton I ever rode, never bested by any of the other British, Italian or Japanese machines who 'wanted a go.'

Nort's 'glory days' were the late '80s and early '90s. He was entered in many national and regional motorcycle shows, garnering a large number of first place awards, including three at International Norton Owners Association Nationals. Those rallies were also the scene of some epic 'road races' on public roads, once described as "illegal as a shotgun murder!" There his reputation was made as one of the bikes to beat. I shall say no more as the statute of limitations may not yet have tolled!

Two years ago Nort went to a new home where he will be cared for, loved, and most importantly ... ridden. The best a proud papa can hope for.

Dave Howe's 'Nort.'

(Motorcycle courtesy Perry Bushong)

R12

I recognize that BMW motorcycles are outstanding, mechanically. They're also so important, historically, that a project of this nature would be incomplete without a suitable representation of BMWs. That being said, I will confess that I've never thought of their engines as being attractive. The BMW motorbikes as a whole have always had their share of style, but the engines strike me as being purely technical and efficient; pretty good ... but not pretty. There, I've said it, at the risk of alienating BMW lovers everywhere.

The 'R12' picture is from a bike rebuilt by Perry Bushong. Perry was a great source of images during the project. This picture was taken at the Up-N-Smoke Fall Car and Bike Show. If anything makes this engine sexy, it's the frame that it's sitting in. Combine the technical engine with that flowing frame and you have the biomechanical lines that I'm looking for.

☙ ~ ❧

Now for a random memory. My brother, Frank, is six years older than me. He started into motorcycling when he was 16 years old with the purchase of a used Ward's Riverside 125 Benelli. Frank said that if he lay down on the tank, and was going downhill with the wind behind him, he could get that 2-stroke to touch 50mph. I believe him, although I never actually saw him do it.

Frank desperately tried to get me involved in motorcycling. But I was 10 years old, and, although I thought motorcycles were cool, my age limited my participation. He'd take me with him on trips, and I would see the world rush by through a yellow bubble shield. When I was 11, he even took me out to a dirt track where he could rent a Yamaha Mini Enduro for me. He really wanted me to be his riding buddy, but the age difference was too great. I have to give him credit for trying, though.

He even gave me a couple of bikes when I was in my mid-teens; opportunities that I wasted. The first was a Benelli 125 Scrambler that only needed the magneto to be replaced. Frank found a salvage engine for me to cannibalize. Unfortunately, he was in the US Air Force by this time and had to return to duty before we could get to the magneto. I was 14 and didn't even know what a magneto was. As far as my mom was concerned, a non-running motorcycle was much safer for me.

Frank didn't give up, however, and, when I was 16 years old he came home on leave and brought me a Honda CL350 to play with. The Honda was ridable, but a little rough, which made it perfect for riding the country trails near our neighborhood – known to the local kids as the 'Pits.' Frank told me that the only problem with the bike was that it had to be push-started. I'd heard of that, but didn't know how it was done.

On a perfect push-starting hill next to our house, Frank explained. Turn on the key, pull in the clutch lever, and put it in second gear. Then, start pushing the bike down the hill and, when I'd got some speed, I was to let out the clutch and give it some gas; should start right up. So, I tried it, and, sure enough, the engine came to life. Sadly, Frank had assumed that I was smart enough to know to jump ON the bike BEFORE letting out the clutch.

I didn't get too badly hurt, but the poor Honda suffered from a bent rear wheel adjustor. Frank had to return to duty, so I had another motorcycle that I couldn't ride and didn't know how to fix. That suited my mother just fine.

(Motorcycle courtesy Mike Cooper. Restored by Perry Bushong)

INDIAN 4

Every time I look at this engine I think; "It's so ugly, it's cute." The styling in this power plant is more architectural than mechanical. Considering it in that vein, then, it's perfect for the project.

ങ ~ ഇ

November 2005, and it's the Thanksgiving holiday. My oldest daughter Elizabeth is home from college, and we're sitting on our back porch of the house just talking about stuff. "I think I want to be famous," I said out of the blue. Elizabeth wasn't surprised with the odd statement, but a little curious. "What do you mean, Daddy?" I told her that I thought I was on the wrong track with selling my engine prints.

Selling prints one at a time is fine, but it's a lot of work for a small return. I should concentrate less on selling prints and focus more on selling myself. I didn't want to be famous for egotistical reasons. It's a simple fact that as an artist, a well-known name carries more value than the picture that it's on. Collectors don't collect pictures – they collect artists.

I couldn't be embarrassed to acknowledge that selling myself, along with the project, was the key to my future. I've been making all this up as I've gone along. Becoming famous was only one more aspect to improvise.

Anyone who knows me well knows about my retirement plan. When I get old and ready to stop working I'll hold up a post office with a pointy stick. The post office part is important because that makes it a federal offence and gets me into a comfortable federal prison. Everyone knows that federal prisons have much better accommodations than state prisons. And I expect to meet a better class of people overall.

Three square meals a day, free clothing, television, reasonable medical care and a place to sleep. I've been to nursing homes – same thing. I haven't been paying taxes all these years for nothing. My daughters don't like the idea and have vowed to take care of me when I'm old. Still ... it's good to have a backup plan.

Maybe becoming famous will be another retirement plan. The question for me now was, what does being famous look like?

MOD M
Indian

INDIAN MODEL 433

In Austin, Texas, there's a group of motorcycle enthusiasts not unlike the NTNOA and the Peckerheads. The Roadrunners British and European Motorcycle Association is more organized than Peckerhead Motorcycle Racing, but they have the same spirit.

Once a year the Roadrunners host a reasonably large vintage bike show at the Ski Shores Waterfront Café on Lake Austin. Phil and I drove down to Austin to attend. Phil was going there to enter his Eye Candy Ducati in the show, and I was going to hunt for engines to photograph.

While the Eye Candy was busy winning whatever category it was entered in, I found this nice Indian 4. I was in luck because the owner was right next to the bike. I asked him if he minded if I photographed the bike. He just said, "Sure, go ahead."

After I finished shooting I wanted to get his name and information about the Indian. But in the five minutes it took me to take the picture, he was gone. I don't mean he had left the show; he just wasn't around the bike anymore. No problem, surely I'd see him around the show. Damn if I didn't get distracted by Mr Budweiser again.

So, this is another engine that I have no owner's name for. Even though I knew that the bike was an Indian, I really didn't know what I had photographed. I even asked some of the Roadrunners if they knew who the owner was. Apparently the bike had come in from out of town, and they weren't familiar with it. Rats!

Jim Bagnard identified the bike for me as a 1933 Indian 4, known as the Model 433. It was just a couple of models away from an Ace, but the distinctive intake manifold identifies it as the Model 433. I was pleased that Jim knew what it was, but I was hoping that I could have titled the picture as 'Indian Ace.' Pity, it just sounded better than 'Indian Model 433.' It turned out to be a popular picture.

While at the bike show I managed to make my most pathetic attempt at selling my prints. I set up five framed prints against a tree trunk with some business cards and a sign with a description of the hat I was wearing. Just in case anyone wanted to look for me ... pitiful.

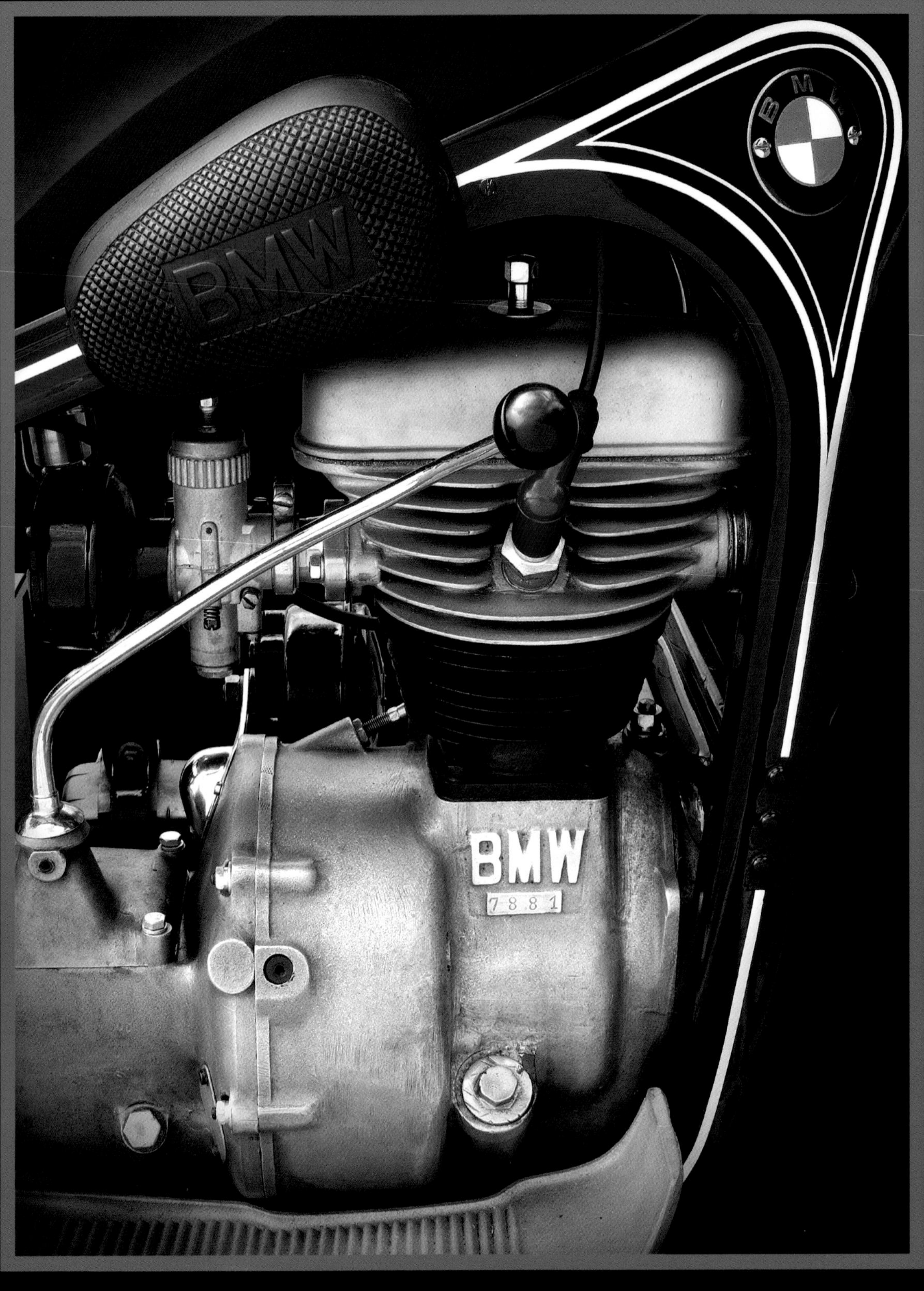

(Motorcycle courtesy Perry Bushong)

With the arrival of 2006 I assigned myself a few goals for the New Year. I needed to find a reliable and inexpensive printing vendor, and I needed to find a better way to sell some prints.

I had been using an inkjet printing service in Utah for about five months. The guy's quality was initially very good, but as time passed ... so did the quality. Nice guy, though, and we had a good business relationship, but one of the rules of professional photography is never deliver a picture that you have to apologize for.

Time again to search for the most elusive beast of all – a reliable printer. I checked online and found a couple of Giclee (French word for high priced inkjet) printers to contact. Commonly, these require a $100.00 setup fee for each image, and $90.00 per print. I was confident in their quality, but with a goal of 50 engines in the project ... well, it was more money than I bothered to calculate. And realistically, these are pictures of motorcycle engines and not precious museum art.

Providence stepped in again with a marketing email from Kodak. I receive their regular professional photographer's online newsletter. January's newsletter introduced the new Endura Metallic paper. Metallic paper sounded good for engine pictures. The web site had a locator for photo labs that were producing Metallic prints. Luckily there was a lab in Dallas called Full Color. If it sent me some good samples, perhaps it could be my salvation.

The samples were dazzling! And the price was in line with what I needed. Not only had I found my reliable and affordable printer, but the Metallic print surface made the engine image jump off the page. Plus, these were real photographic prints, and as an old get-your-hands-wet photographer I preferred them to inkjet. I could sell and print on demand without keeping any inventory. I was a happy man.

☙ ~ ❧

The R2 comes from BMW's early years when it was still experimenting with different engine configurations. The R2 isn't common or popular, but it's different, and has the trademark no-nonsense engine conjoined with a fluid frame; technically elegant.

(Motorcycle courtesy Jeff Dancey)

SQUARIEL

Here is another nifty engine I found at a Lake O' The Pines Rally. The Ariel Square Four was always near the top of my 'must have' list. This fine example belongs to Jeff Dancey of Houston. Jeff was at the rally with other members of the British Motorcycle Owners Association. Not only was Jeff pleased to have me photograph his Ariel; he even gave me a couple of beers to enjoy while I did it.

☙ ~ ❧

Having a reliable, affordable and high quality print product, the Metallic prints opened up my sales options. And those options could help me with my ultimate goal of becoming famous. I'd like to say that I wanted to be more famous than the Beatles, but that would be blasphemous. Besides, I'm only making pictures of engines.

So, since I was playing famous artist, I decided to find an art dealer to represent me. There are a lot of motorsports art dealers out there if you look for them. I sent out a half dozen inquiries to see if there was any interest in representing my work.

I found that many of the motorsports art dealers only carry Harley-Davidson related art. My engine pictures were practically invisible for this group and I never heard back from any of them. Luckily, I did hear back from Mike Mayer of CruisinGoods.com. His operation is web-based, and he carries a wide variety of motorsports art, including motorcycles, autos and aircraft.

I confessed to Mike that I had no idea what I was doing, and that I would need his help to get started. After several email exchanges we worked out that he would list my work on his site and, when he received an order, he would send me the information and a check for my share of the money. That plan has worked out pretty well.

I now could say that I was represented by an art dealer. Cool. That helped to give my work a little more credibility. And the inclusion of my name among his list of other well-established motorsports artists didn't hurt either.

We didn't have many sales initially. However, by Christmas 2006 we had some satisfying sales. I still didn't feel like a real artist, but at least I was beginning to think like one.

Brough Superior
"SS 80" Model
Brough Superior

SS 80 MATCHLESS SUPERIOR

Up until this point the production of engine images has, more or less, corresponded with the chronology of the project's story. The images were created at about the same rate as the story was unfolding. That changes at this stage. Once I had a pool of images to draw from, I completed the pictures as I found the time, but the story kept rolling right along.

ଓ ~ ଛ

This Brough Superior belongs to Gene Brown and was restored by Brough and Vincent historian Herb Harris. I photographed it as it sat in the Motorcycle Classics' European People's Choice Concours d'Élegance at the 2006 Barber Vintage Festival (more about that event shortly).

ଓ ~ ଛ

Spring of 2006; Arnold Barnhart asked me how much it would cost him for a 20in x 24in Metallic print of the Eye Candy. Arnold is a buddy, a business associate, and a Peckerhead, so I told him 50 bucks. He said, "Fine, make me one." The request seemed a bit strange, because Arnold is a big motorcycle enthusiast, but he didn't have any particular affinity for Ducatis.

I was pleased that I sold a print, but I had to ask, "What do you want a print of the Eye Candy for?" Arnold replied with an air of matter-of-fact confidence, "I like the picture. I'm going to send it to Jay Leno as a gift." "Really?" I almost laughed. "You know Jay Leno?" Still matter-of-factly, Arnold said, "No, but I know the guy who runs his garage and he'll give it to Jay."

This was typical Arnold. He's one of those people who has apparently done everything. And damned if he doesn't consistently back up what he says. I first met Arnold shortly after starting Trick Photography. He found my business card at Up-N-Smoke and called me to photograph a motorcycle for him. He was the new US distributor of the Moto Guzzi powered Ghezzi & Brian sport bike. After the job was over I invited Arnold to meet the Peckerheads. He's been a buddy and a Peckerhead ever since.

A few months later at a regular Friday night Peckerhead meet, Arnold told me that he had spoken to his guy and that the Eye Candy print was framed and hanging in Jay Leno's garage facility. Now, that was cool! ... but. "Really?" I said with skepticism. "I mean, is it REALLY hanging in Leno's garage?" I was now a little concerned. "Yeah," Arnold said nonchalantly. Then he tried to tell me exactly where it was hanging in which hallway. "Okay Arnold, this has to be no bull. If I tell people that Jay Leno has one of my prints and it turns out to just be a story it'll make me look bad." I was serious; and that's unusual for me. Arnold looked up from his Guinness and just said, "I understand. It's there." Wow. Jay Leno has one of my engine prints ... as far as I know.

"Tonight's guests are Madonna, and engine photographer Daniel Peirce ..." Delusional, but fun to entertain. I guess I owe Arnold a Guinness or two.

(Motorcycle courtesy Mike Thomson)

BLUE STAR

I kept running into Mike Thomson and his impressive bike collection. This is his BSA Blue Star that I found in an Antique Motorcycle Club of America bike show. It always seems that BSA and I are on the same page when it comes to engine styling. The Blue Star has a stand-alone personality. It's a living entity that BSA managed to capture in a motorcycle frame.

☙ ~ ❧

Dave Howe pointed out an ad in a motorcycle magazine for a little company that put photo images on ceramic tiles and coffee mugs. And they specialized in motorcycle subjects. Dave thought I might want to expand my product offerings. Perhaps I would, I hadn't considered it before. I took a look at their web site.

Their primary product was dye-sublimated multiple tiles arranged to make large tiled mosaic picture. I didn't like the idea of the engines being segmented. But they also made single tiles with easel backs attached. That would make a neat stand-alone picture, perfect for a desk or bookshelf. That started me thinking again. I could sell unique and less expensive versions of my engine pictures. Tiles and mugs could be the beginning of that. I sent the vendor an image file of the 'Chief' picture as a test. I explained what I was planning and that we had an opportunity to establish a potentially profitable business relationship.

Establishing a good business relationship is important to me. I like to think of my business associates as friends. And I'm loyal to my friends. They can trust me to deal with them with integrity, and I trust them to deal with me the same way. It's very important to me. It's part of the Texan culture we call "doin' bidness."

The samples they sent back to me were excellent. I showed them around to other artists – and the Peckerheads. All agreed that they would be worthy and viable products. I didn't have money to carry an inventory so the vendor agreed to produce in short batches as I needed them. Great, we were in "bidness."

The tile folks were also motorcycle people, so our relationship started on a good foundation. They even sold some of my images on tiles when they set up a booth at motorcycle shows. We had a good time selling a few mugs and tiles. Even the Barber Museum bought some to sell in its gift shop.

Alas, all good things come to an end. Slowly, their quality began to deteriorate, until after about six months I received a batch with color so bad that I considered it unsalable. I tried to contact them about it, but it was too late. They had gone out of business. I never found out exactly why. They just disappeared. It would take me quite a while to find a vendor to take their place.

(Motorcycle courtesy Gene Brown)

GOLD STAR

A strange thing happened. I was wandering the antique bike shows at the 2006 Barber Vintage Festival when I spotted this lovely Gold Star. I assumed the man polishing on the bike was the owner and approached him. "Hi, are you the owner of this Gold Star?" "I am," he said proudly as he looked up from his polishing. I began my standard introduction speech, "Hi, I'm Daniel Peirce and I'm working on a photography project about ... " He interrupted, smiled, and said, "I know who you are."

Gene Brown had already heard of me and of the project. Maybe this famous thing had a chance. Gene was pleased that I wanted his Gold Star engine for the project, and suggested that I also photograph his SS80 Brough Superior.

ꙮ ~ ꙮ

May 2006, I was busy making images, and I now had an art dealer representing me, but I still wasn't selling much of anything. I had read an article about artists selling their works on eBay. Hmmm. I had dabbled with eBay but I never took the time to really learn about it. And taking the time to learn about it still didn't have much appeal for me. But maybe it was time to bite the bullet.

After about a week of research, setting up an eBay account and setting up with Paypal, I decided to start building an eBay store. I was actually dreading all this to some degree. This was because I knew that once I got involved in it I would attack it with the same obsessive focus that seems to afflict me so often. Damn – my fears were well founded.

Frankly, I was giddy with my first sale. I only made about $10.00 on the sale because I started my prices very low. Still, I was amazed that eBay could be a viable outlet for my art. Damn it ... the obsession was kicking in. Soon I was tracking sales, experimenting with different formats and studying hit trends. All while I was still supposed to be rebuilding my CB750 engine. What engine?

But the prints started to sell. And they were selling worldwide. I had customers in the UK, Germany, France, Hungary, Spain, Japan, Brazil and Australia. Even sales to exotic places like Canada and Wisconsin. I still wasn't making much money, but I was gaining a reputation, and that was becoming a major part of my plans.

(Motorcycle courtesy Jim Turner)

Z1

Early in the project I photographed Johnny Cool's Z1 as it sat in a vintage Japanese bike show at Honda Suzuki North in Dallas. I really enjoyed the show – mainly because my Landshark won a trophy for best vintage modified. I wanted a picture of a Z1 because of its historical significance as one of the first commercially successful dual overhead cam four bangers. It was also the bike that seriously challenged Honda's dominance in super bikes.

As I was shooting the Kawasaki I was having trouble finding the trait that made the engine distinctive. Of course, it's the double lobes at the top of the heads that house the camshafts. I tried various angles but I didn't come away very excited about what I saw through the camera lens. However, I've been surprised before with pictures that had benefited from the retouching process.

Well, sometimes the magic works, and sometimes it doesn't. I managed to make a picture out of it, but it never satisfied me. The image was added to the collection because the project was still young and needed populating. Besides, I kind of thought the picture made the engine look like a robotic face. I resolved to keep my eye out for another Z1 and try again in the future.

~

At the 2006 Barber Vintage Festival I found this very clean Z1 in the Vintage Japanese Motorcycle Club display. I met the bike's owner Jim Turner and he was happy to let me shoot it. By this time I had developed a somewhat consistent look for the engines. Turner helped by holding the reflector for me and the session went smoothly.

The finished picture pleased me and I replaced the first Z1 image. Feedback on the picture told me that Z1 fans like the image, but they would prefer that it wasn't shown with the engine guard. I can see their point; still, it's a good picture of a great engine. Perhaps I'll keep looking.

The original, but discarded, Z1 picture.

(Motorcycle courtesy Phil Dansby)

900 SS

This is another one from Phil Dansby's collection. Phil bought this Ducati Desmo Super Sport new from Doc Storm in 1979. It's a sexy engine in anyone's book, but the entire bike is a work of engineering art.

I first photographed Phil's 900 SS in 2001 for the second article in the Texas Roadrunners series in *Ride Texas Magazine*. Dave Howe wrote the article and titled it 'Rombo di Tuono, Y'all!' Translated from Italian it means, 'The Rumble of Thunder, Y'all!' The opening shot for the piece featured the bike with Phil's Up-N-Smoke BBQ Restaurant in the background.

A special feature added to this Desmo is the Gear Gazer. That's the unique little window over the bevel gears that shows oil flow when the engine gets to temperature.

~

September 2006. Jim Bagnard called me and asked if I wanted to share a couple of vendor spaces at the Mid-America Antique Motorcycle Auction in Dallas. I had just finished constructing a large stand to display my prints at the Barber Vintage Festival in October. This would give me a good chance to give the display a test run. Didn't cost too much, so I accepted.

Turned out that we were the only vendors there. Vendors add interest to an event like this so the auction organizers appreciated us being there. Jim was selling his motorcycle lift tables and I was set up with a display of a dozen Metallic prints and my litho posters. As always, I didn't care too much if I sold anything or not. I had my traveling engine exhibit and I enjoyed talking to folks who appreciated the project.

I sold enough to pay for my effort and made some valuable contacts as well. One of the people I met was John Furge. He asked me why I didn't have a Yamaha XS-650 in the collection. I told him that I knew they had a cult following and that I had been looking for one, but hadn't run across any yet. "No problem," said Furge. "I have four of them."

John lived locally and we made loose arrangements to get together so I could photograph one of his engines and his entire XS-650 collection. Cool.

This is the picture of Phil's Ducati 900 SS we used in the opening of the Texas Roadrunners article in* Ride Texas *magazine.

(Motorcycle courtesy John Furge)

XS-650

Who knows where obsessions come from? One day John Furge of Plano, Texas, decided he wanted a Yamaha XS-650 like he knew when he was a kid. That idea started an obsession that led him to owning four XS-650s.

The day we shot this XS-650 engine picture, we also photographed his entire collection of bikes. I told John that if he could write 600 to 1000 words about them I thought we could get a magazine article published. Turned out that writing is another of John's obsessions. Eight months later the article and photos were printed in the Vintage Japanese Motorcycle Club magazine. We even made the cover.

Since Furge has his own way with words and a special perspective on the XS-650, I asked him to share them in this book. From John Furge:

This beautiful 653cc, 2-cylinder, 4-stroke was born in 1970 as the power plant for the Yamaha XS-1, and was produced every year until 1984. The engine in this picture is a 1973 TX-650, and produces 53bhp at 7000rpm. Like the 1972 model XS-2, it has a compression release that operates in conjunction with the electric starter. Most owners ignore the electric start and use the kick-starter. The compression release disappeared after 1973.

A milestone in Yamaha's history, this is first 4-cycle engine Yamaha produced as all models heretofore were 2-cycle engines. The market for Japanese bikes had been the small displacement model, but they began to compete with American and British made machines. The XS-650 was styled much like the classic British bikes – BSA and Triumph, and was intended to compete in that market. At the same time, Honda introduced the 4-cylinder across CB750, and the race for displacement was on.

The 650-twin model of Yamaha motorcycles has a large following. One can find many sites on the Internet by searching for 'XS-650' on Google.

Postscript: What makes a car, motorcycle or engine look 'tough' or attractive?

Our human minds possess a visual filter that can be referred to as a 'dimension modulator.' Certain patterns and size relationships are good ... the others are not. This ultimately goes back to ancestral days, and is an evolutionary tool to assure survival of the fittest. What is properly laid out is unconsciously attractive. Take beer commercials for example – see the girl ... Okay, that's an example of beauty. Deviations there from become ... well ... less beautiful.

Mathematically speaking, the term 'Golden Ratio' is the underlying cause. It's actually hard wired into your brain. Any three dimensional item upon which you gaze is measured for the ratio being echoed thorough its structure. The Golden Ratio is a length of rope, which, if cut makes two pieces of rope, the shorter which is ½ the length of the longer. This is approximately 1.6180339887 to 1.000.

Nature frequently echoes this ratio, architects do likewise, and, of course, so do engine designers. Of all the engines you see in this book, consider symmetry and the ratio of chunks of metal (the head size to the casing size, for example)

A way to remember this concept; look at your 1st finger tip which is equal to 1, and the finger phalange just below it is 1.6, then the 2nd phalange equals one, and the 3rd phalange is 1.68. If you study this long enough you will begin to see it run throughout the human body.

For further study search for 'Golden ratio' and 'Fibonacci numbers.'

John Furge and his friends.

(Motorcycle courtesy George Tuttle)

TRIUMPH T-120R

It was time to add another Triumph to the collection. Fellow Peckerhead, George Tuttle, volunteered his pristine T-120R to the cause. I photographed it in his driveway on a Texas summer Sunday morning before it got too hot. Too late; it was already too hot.

ᴓ ~ ᴓ

Anticipating in the spring of 2007 that I'd soon be ready to produce a book, I came up with a great idea. I would send out a call for motorcycle enthusiasts to contribute essays about the engines for the book. That way I'd have a variety of impressions to go with each engine picture. That idea turned out to be a dismal failure. Seems there aren't as many folks who like to write as I had anticipated.

I received only a couple of contributions. The first one I received was from Christian Clarke. So, to honor Christian's efforts I'm printing a portion of his essay on the T-120R:

"The first Triumph T-120 Bonneville was offered for the 1959 season with 649cc, a splayed port cylinder head, and twin carburetors. The 'T' stood for Tiger, the '120' advertised its alleged top speed, and 'Bonneville' was just a nickname – referring to Johnny Allen's Salt Flat record – no one thought would stick. The nickname took and the motorcycle industry was caught gasping for breath and bandying for second best.

Throughout the Sixties, changes to each model year helped refine the peaky nature of the Bonneville's high performance. By 1969 Triumph had all its ducks in a row. Malcolm Uphill won that year's Isle of Man Production TT on a Bonneville, averaging 99.99mph per lap, with the first ever 100mph lap by a production motorcycle. Obviously, the race winning production bikes were a little bit different than the showroom floor models.

"Restorations for classic motorcycles generally fall into two categories, show bikes and riders. Trailer Queens are seldom ridden because British bikes leak. They've never not leaked. Once oil is pumped throughout the bike, it's a trying task to keep it in its place. Gas drips from the carbs, chain-oil coats the rear wheel, and over-tightened bolts provide myriad possibilities for escaping oil. And there's lots of oil. The 1969 Bonnie uses four different types of oil: engine oil, gearbox oil, primary case oil, and fork oil. It is not uncommon to see different shades of lubrication collecting underneath a parked Triumph.

"Triumphs in the late '60s, unlike their Japanese counterparts, required attention. Before every ride, bolts needed to be tightened, fluids needed to be checked. Routine service was a weekly endeavor, not just every 3000 miles. By the time a British bike had passed from one owner to the next, oil changes and tune-ups were rarely given their proper due. Cut to 30 years later and you have a lot of leaky, foul-minded motorcycles, and owners craving more civilized machines. You don't get into old bikes without an idea of their maintenance needs."

(Motorcycle courtesy Barber Vintage Motorcycle Museum)

ASTRO 360

As a teenager in the '70s, and my buddies into dirt bikes, I often heard about Bultaco. The name was always brought up as a revered mythological hero. Indeed, as a legend, because although the guys talked about them, I don't think that any of us actually ever saw one. I just thought it had a funny name.

The Astro 360 was named after the Houston Astrodome. The Astrodome was a popular venue for flat track racing in the 1970s.

☙ ~ ❧

I'd been hearing about a new vintage motorcycle magazine called *Motorcycle Classics*. The US vintage crowd had been without a serious periodical for quite a while. It was good to see that *Motorcycle Classics* had taken up the cause with a quality publication.

I sent the editor, Richard Backus, one of my new photo tiles, with a press release to introduce myself. If this magazine caught on it would be serving the same audience that appreciated my engine project. I wanted this magazine to survive.

I didn't have an advertising budget – mainly because I didn't have any budgets at all. But now that I was selling through an art dealer and eBay, perhaps a little ad wouldn't hurt. I contacted the *Motorcycle Classics* ad salesman and bought a little 2 inch square color ad that would appear in four consecutive issues. A big expense for a little business.

☙ ~ ❧

The CB750 engine was completely apart and scattered across my garage worktable. The 'month-long' rebuild was going on its eighth month. Work on it had been sporadic with everything else that was going on. I had all the parts I needed now and I was tired of being without a bike. It was time to switch obsessions.

Randel Bird, a fellow Peckerhead, came over to the house several times to help me get the engine back together and back into the frame. Randel is a good mechanic, and his only requirements were a good stock of beer in the fridge and the radio tuned to Cajun music. The rebuild finished up well. But twisting that heavy in-line four engine back into the frame was hard enough to convince me that one engine build per lifetime is enough.

Sport
DUCATI
DUCATI

750 SS

Ducati's timeless masterpiece, the 750 Super Sport. If you can't see beauty in this engine, then you're holding the wrong book. The 750 SS is a shining example of artfully combining the symmetrical and the asymmetrical. No wonder this bike is so highly sought-after by collectors.

☙ ~ ❧

Summer of 2006. By now I was selling about five prints a month and making about $10.00 profit on each. Not much of a profit, but I had to start somewhere. Everything I was making was going right back into the project. I wasn't making a profit and I was operating at a loss. I think I mentioned that I'm not much of a businessman ...

To keep Trick Photography afloat, I was still accepting the occasional magazine assignment. In June, with the recommendation from the folks at *Ride Texas*, I picked up a plumb assignment from *Texas Highways*. *Texas Highways* is a popular Texas travel magazine published by the state department of transportation. It always uses beautiful photography, and I've always wanted to work with it since I started in the business.

I was hooked up with freelance writer Dale Weisman, who had also pitched the idea to the magazine. The article was to be about motorcycling in the Texas hill country (the only area in Texas with an abundance of curvy roads). *Texas Highways* had contacted *Ride Texas* looking for a photographer who knew motorcycles. That turned out to be me.

The only problem was that my bike's engine was still in pieces. In desperation I asked Jim Bagnard if I could borrow one of his bikes for the trip. Without a blink, Jim offered me his Kawasaki Drifter 1500. Only a true friend, and a true Peckerhead, would be so generous.

Dale and I met in Wimberley, in central Texas, to begin the assignment. From there we started traveling and visiting towns with names like Blanco, Medina, Fredericksburg, Leaky, and Luckenbach. And we rode on roads with names like The Devil's Backbone and The Three Twisted Sisters. It was while riding on one of the technically demanding Twisted Sister roads of Hwy 337 that I suddenly realized I was straddling a borrowed cruiser on a peg-scraping country blacktop. Was I insane or just naïve? Apparently both.

I took great pains to make sure that the Drifter always remained upright. The bike performed flawlessly, and I put 1100 miles on it in three days. I gratefully returned it to Jim – washed and polished, of course. It was quite a trip.

Half of the money from that assignment went to Trick Photography, and the other half went to Pam. My wife puts up with my obsessions and I have to reward her when I can. *Texas Highways* was pleased with the photography and Dale's writing. One of my pictures even made the back cover – a prized spot and a great ego booster for me. As if I need an ego booster.

(Motorcycle courtesy Barber Vintage Motorsports Museum)

MANX

Phil's primary motorcycle passion has always been with Nortons. He was even one of the founding members of the North Texas Norton Owners Association. He made a special request for a picture of the Norton Manx engine. For Phil, the Manx is the epitome of the sexy engine.

The Manx became a staple for the privateer racers. Ready to race right out of the box. Just add body parts to suit your type of racing, and you were on the track. To me the engine just looks powerful ... like a flexed muscle.

☙ ~ ❧

Since the Landshark was up and running again I started riding it to work every day to break it in. A small tapping sound started under the valve cover on the right side after only 50 miles. After 100 miles, I was riding home from work when the tapping became noticeably louder, and the engine belched oil onto my right pant leg.

Both of these events were disconcerting, but the oil belch was especially troubling, as I could find no obvious opening for the oil to leave the engine. I was smart enough to realize that something was seriously wrong and parked the bike back on the lift table. The truly unfortunate part of this was that on the CB750 SOHC the valve cover couldn't be removed without removing the entire engine from the frame. I'd have to wait a while for the disappointment to subside before I'd be able to bring myself to wrestle that engine out of the frame again.

☙ ~ ❧

It was the beginning of October and time for the 2006 Lake O' The Pines Rally. Jim and I heard that vendors would get in for free this year. I like free. We decided that this year we would attend the rally as official vendors. Jim loaded up his flat bed trailer with four lift tables along with his Sportster and Drifter for our riding pleasure. I loaded up my new display and prints in his pick-up and we were off to the rally.

At the rally we dutifully manned our booths. But we were mainly there to screw around, drink beer and enjoy the rally. We both managed to sell just enough to pay for expenses, which was becoming a standard operating procedure for me. If you can't have fun doing what you're doing, then why bother? Good thing I have a steady day job. I don't think the rally is letting vendors in for free anymore.

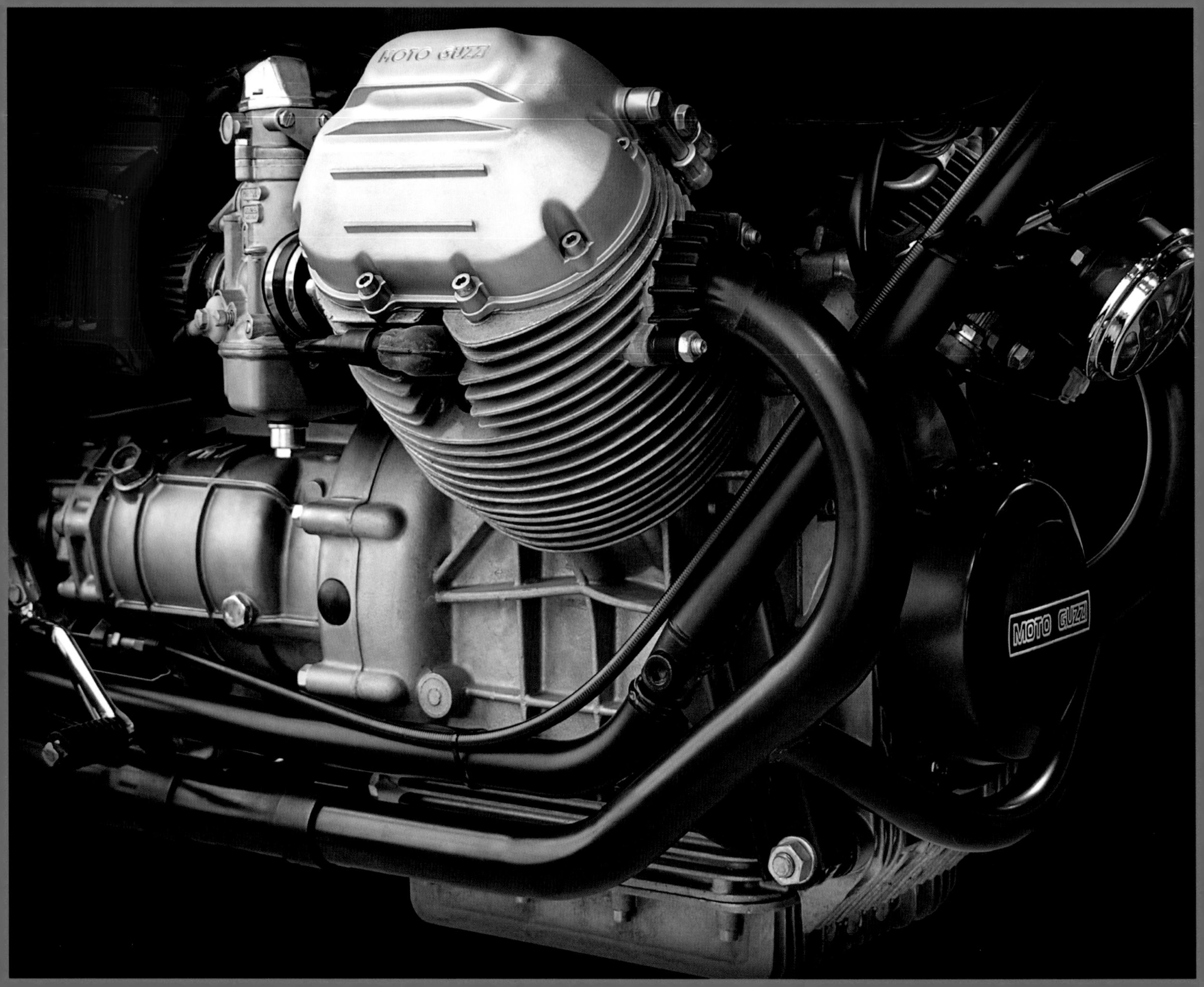
MOTO GUZZI
MOTO GUZZI

LE MANS

Surprisingly to me, the Le Mans print was an instant hit. I think this is partly because this is a very good picture of a very good engine, but mostly because there is a large group of fanatical Guzzi-heads in the world. And maybe those Guzzi-heads appreciated that someone thought this engine was sexy enough to deserve a place in the collection. This one is for you Moto Guzzi fans.

☙ ~ ❧

October 2006, and Jim and Randel and I were in Leeds, Alabama, to set up at the 2nd Annual Barber Vintage Festival. Randel had recently started Gearhead Designs as a side business, making lamps out of discarded engine parts. Jim brought eight lift tables to sell and I had my new display with a dozen Metallic prints to show.

It had rained hard before we arrived, and our designated swap meet spaces were knee-deep in mud. They assigned us new spaces on higher ground that worked to our advantage with parking our heavy trailer. To save money we all decided to camp in Jim's 40ft enclosed utility trailer during our time at the festival.

The Friday and Saturday of the event saw sunny skies and warm weather. The crowds were decent considering it was only the second year of the festival. Jim sold all eight of the lifts he brought. I wrote enough print orders to pay for my trip. I was still putting out a lot of effort without making a lot of profit. At least I was having fun doing it.

Fortunately, I was able to photograph several engines while at the festival. I found the Gold Star, the Blue Star, the Brough SS 80, and the Z1. I also shot a few more that haven't been made into finished pictures yet. The Barber Museum was too crowded during the festival to even think about photographing in. Jim and I still went to the museum to look around. While there, I ran into Jeff Ray and told him I'd contact him in the spring to arrange permission for another photo session at the museum. I'm not sure he remembered who I was.

Saturday night a cold front came through with dropping temperatures and bringing another rainy deluge. It was good to be in the watertight trailer. We woke up early Sunday morning to standing water, collapsed tents and a cold soggy turf. It didn't take a crystal ball to foresee the muddy mess that was coming when all the trucks and trailers started moving out. We decided to forget the Sunday session and pack up and move out as quickly as possible. We barely got that 40ft trailer out of the field and we weren't even going through the mud. We left the park at 7:30am. Later, we heard nightmarish stories from folks who waited until later to leave. Most of the tales involved cables and tow trucks.

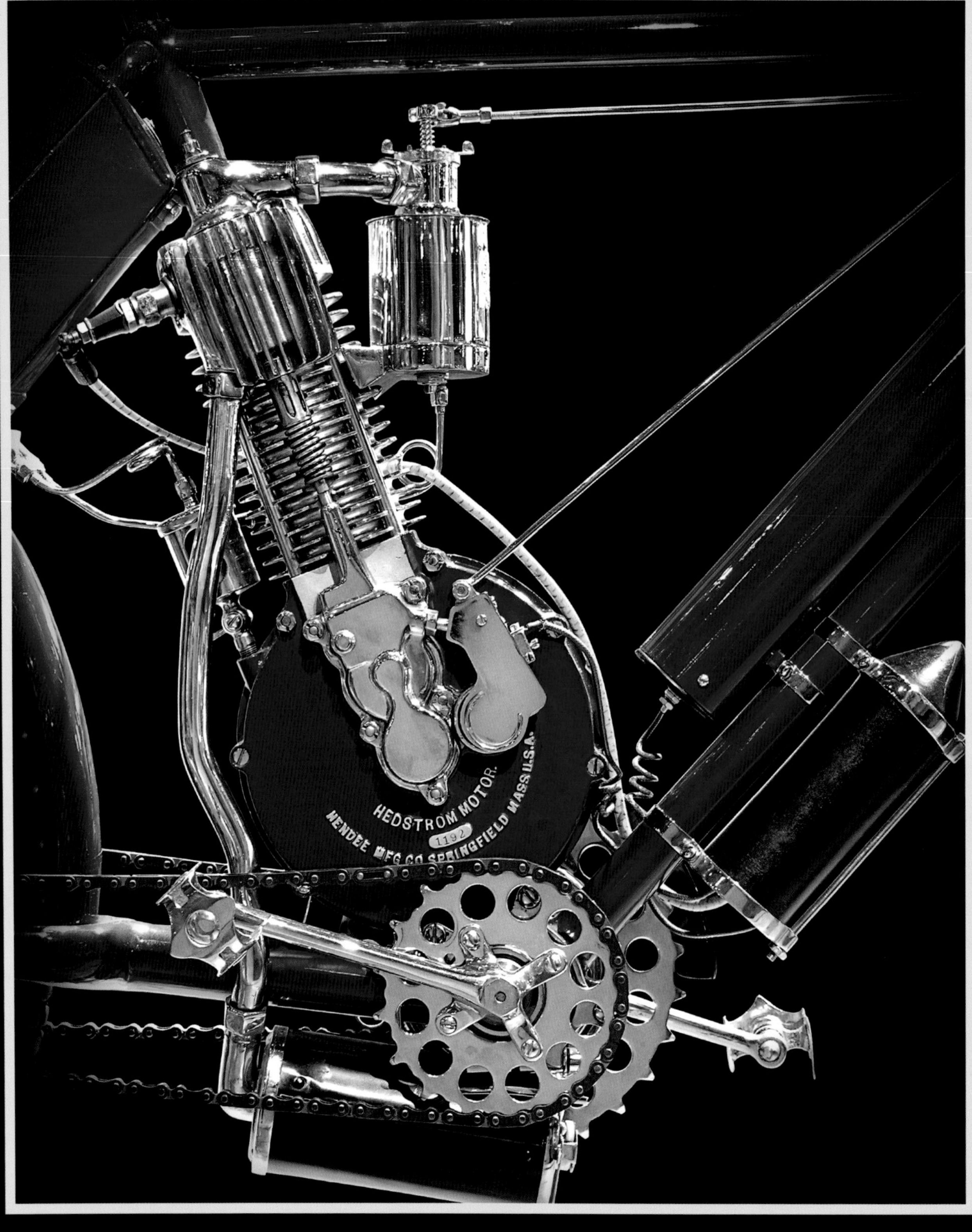

(Motorcycle courtesy Barber Vintage Motorsports Museum)

CAMELBACK

Although it's strange to title this picture 'Camelback,' because you can't see the distinctive humped fuel tank that gives the motorbike its nickname, it's a famous Indian motorcycle with a famous name; so what else could I title it?

☙ ~ ❧

I sent out a new press release in October 2006 to a few consumer motorcycle magazines and web sites. David Edwards of *Cycle World* was kind enough to do a write up on the project, both on the *Cycle World* web site and in the magazine.

The web article was titled 'Pornography for Gearheads,' and spoke about the project and the Peckerheads. Edwards had experienced the Peckerheads first hand. He also included a number of images from my web site. I hadn't seen the magazine article because the issue hadn't been distributed yet.

In December, my half-brother Rob contacted me to tell me that my father was dying. My dad had a bad liver and a bad heart. They couldn't fix one because of the weakness of the other. After a major heart attack the doctors told him to start getting his affairs in order. His liver would slowly shut down until he would go to sleep and never wake up. He went home to be with his family and wait for the end.

My half-sister Michele arranged and paid for airline tickets so my sister Nancy, her daughter Michaela and I could go to Florida to see him. Although I love and like my dad we've never been close. My parents divorced when I was six, and correspondence and visits between my dad and I were few and far between. No real reason, it just worked out that way. I didn't even know he was living in Florida until Rob called.

As we waited in the airport terminal for our flight I visited a magazine shop. On the shelf was the *Cycle World* issue with my article in it. It was the first time I had seen it and I purchased two copies.

When we got to see my father he was in his living room. He was obviously ill, but his spirits were good. In his pragmatic approach to life (and death) and his sense of humor, my dad and I are identical. You just can't keep a Peirce down. Death was just an unavoidable inconvenience. Dad was smiling and cracking jokes in between coughs. The same attitude I would have.

I brought out the magazine and showed it to my dad. He carefully read the article, looked up, and said, "You're famous. Joyce, my son is famous. I'm proud of you." It was deeply satisfying to be able to bring the article for my dad. My success would be his as well. But he would have been proud of me even if I were a mass murderer; as long as I was a good one.

We could only stay for three days, and before we left he told me not to bother coming to the funeral. Our visit while he was still alive was all he needed. He was a pragmatist to the end. A month later Rob called to tell me that Frank Robert Peirce, Sr had passed away.

(Motorcycle courtesy Barber Vintage Motorsports Museum)

The CBX engine was on my 'must have' list, and was also one of the most requested from fans of the project. My next-door neighbor has a nice '79 CBX with the extra sexy DG 6-into-1 exhaust. I wanted a CBX that was closer to stock, though – for some reason.

This one came from the Barber Vintage Motorsports Museum. It sits on the second level of a tiered display, which made it about head high. I liked the black paint treatment, even though I wasn't sure if it was stock or not. This wasn't the angle I would have chosen, but it was the only one I could get. I thought that maybe it would look artsy shot from underneath.

The resulting image is a good picture of a CBX engine, but it's not a great one. I've been called on it more than once from CBX fans. Some don't like the black pipes, and some don't like the angle. Some don't like both. My honest opinion is that I agree with them. It's good for the collection for now, but it will likely be replaced some day.

ଓ ~ ଓ

The Christmas season 2006 had steady sales, and between my eBay store and Cruisin' Goods I probably sold twenty prints between November 1 and December 15. Before then I was selling about five prints a month. I considered the Christmas season to be a great success.

I began to notice a sales trend. There were the really famous engines that consistently sold well, such as the Black Shadow and the Brough Superior, but the best selling group of engines came from motorcycles popular in the 1960s and 1970s. I determined that those were the machines that older motorcyclists still had fond memories of. I had to include myself in that group. So, for the sake of print sales I needed to make sure that I had several engines from that period in the collection. On the other hand, this was still an art project about all of the sexy motorcycle engines since the beginning of the motorbike. I had to be careful not to only photograph engines that would sell. For every engine from the '60s and '70s, I would have to make sure there was at least one photographed just because it was a cool looking motor. After all, I was an artist first, and a businessman somewhere around last.

(Motorcycle courtesy Barber Vintage Motorcycle Museum)

K.T.T. MK VIII

Velocette made elegant and complex engines. The triple oil manifold feeding the heads makes the K.T.T. Mk VIII pretty distinctive. I'm still looking to make a picture of the Thruxton.

☙ ~ ❧

January 2007; with the New Year came the time to set new goals. I made some ambitious ones. In the coming year I wanted to set up my web site so it could accept online orders, have an exhibit of engine prints, and find a publisher for a book.

I'd had my web site since I started Trick Photography in 2001, but it was mainly geared to be an online portfolio. George Tuttle had taught me just enough FrontPage to make a functional web site. Trickphotog.com worked, but it certainly wasn't fancy. Entering the world of e-commerce was something I had resisted for a long time. I gave myself a deadline of February 1 to have an operating shopping cart system.

Having an exhibit was a personal goal. Having been a commercial photographer most of my career had left me few opportunities to create art. Consequently, I never thought of myself as an artist. An art exhibit would help validate that the project was more about photographic art than a commercial enterprise. The only thing I knew about having an exhibit was that it was going to be expensive. Everything else I would just have to make up as I went along. Just another adventure. Now ... where to hold it?

When I first started the engine project, I contacted a few publishers about the possibility of producing a book. But the project was young and the publishers would have to imagine what a complete picture collection would look like. I never heard back from any of the publishers I queried. I learned that book publishers need either a completed manuscript or a fully conceived idea. Otherwise, I was merely wasting their time.

I calculated that I'd reach my fifty-engine goal in the fall of 2007 and I would start contacting publishers then. So, with goals determined and a time line set, I looked forward to see what 2007 would bring. Oh yeah, then I realized I'd left out one goal. I still had to get the Landshark running again.

(Motorcycle courtesy Barber Vintage Motorsports Museum)

F.B. MONDIAL 1957

Although I love the overall look of the Mondial, I think it looks like it was pounded into shape by a blacksmith. This was a champion racing bike, and all the parts on the engine are there because they need to be. Still, it has a raw mechanical esthetic to it. Biomechanically, it's practically feral.

☙ ~ ❧

George Tuttle is a smart guy. He had to be to fly F-4s and F-16s in the Air Force. George is the web master for the NTNOA, and the guy I go to for answers to my web questions. He gave me all the information I needed to transform my plain web site into a plain e-commerce web site. Some day I'm going to learn enough to make it into a fancy web site. George even came over to my house one evening and helped me fine-tune all the settings. As the song says, "I get by with a little help from my friends." Even if all my friends are Peckerheads.

I liked eBay but it tends to whittle away at my profits. With store fees, insertion fees, final value fees, and the Paypal fees, I figure they eat up 15% of my profit. Having sales through my own web site would still have expenses, but at least I could present my entire engine collection without chewing up all of my return. I kept my eBay store up and active because it gave me exposure in the market, as well as a sale now and then.

After my semi-annual web site revamp I worked on setting up the shopping cart system. To my complete surprise it actually worked. Hours of learning and inputting had paid off, and I made my February 1 deadline.

☙ ~ ❧

The prior summer I had been on location in Arizona shooting a Firstgear catalog. I got into a conversation with one of the guys who was riding motorcycles for us, and he told me that he used to drag race Honda CB750s in the 1970s. He told me that they were great engines for racing, but they had a design flaw. Tech inspections at the drag strip required an inspection of the lobes on the camshaft. But you can't remove the valve cover without pulling the engine out of the frame ... bad. The culprit was a pair of frame struts that strengthened the spine. It made a stronger frame, but at a big cost. The drag racers simply cut a section of the frame struts large enough to remove the valve cover. Then they would replace the sections and secure them with ring clamps. Since the bike was essentially running in a straight line for a drag race, the weakened frame struts had minimal effect on stability.

Remembering that story got me thinking about my own valve cover problem. This was going to take some careful planning.

(Motorcycle courtesy Barber Vintage Motorcycle Museum)

MIRAGE

If you're a Laverda fan, and you don't recognize this particular model, don't worry, it's fairly rare. This 1200 Mirage was a special Laverda anniversary model. Not many were made, and very few of them made it out of Europe.

I originally titled this picture as 'Jota 1200.' I was quickly and emphatically corrected. I learned that it is NOT a Jota 1200. I also learned to do a little more research before I titled my pictures.

ᘓ ~ ᘔ

Around the Peckerhead table I was getting the same looks I did when I announced that I was going to rebuild my engine. John Buckles was looking at me like I was crazy, and he asked me flatly, "Are you crazy? You're going to CUT your frame?"

All I was doing was trying to get suggestions for making the frame sections reattach securely. I had my own ideas, but I had a wealth of mechanical knowledge sitting at the table. Some guys just shook their heads in pity. Some agreed that I'd ruin the bike and destroy its cornering stability. But a few guys actually gave the problem some consideration.

I was aware of the risk I would be taking by modifying the frame. But the thought of pulling that engine out again just to look under the valve cover seemed ridiculous. To me, the frame modification seemed less ridiculous. I wanted to collect design suggestions before I settled on a plan.

Jim started sketching ideas on a napkin. He's a pilot by trade, but he has a degree in mechanical engineering. Peckerhead Steve Ledbetter offered his well-equipped machine shop and his machining skills to the task as well. I was still going to give myself a couple of weeks to think about it before I settled on a design.

ᘓ ~ ᘔ

While looking for a new mug and tile vendor I happened across a company that sold self-adhesive blank photo magnets. I ordered a small amount to see what I could do with them. I made 4in x 5in Metallic prints and assembled them as photo magnets. They were beautiful. The Metallic paper looked good as a large print, but they looked even better as small prints. They just had to be seen in person to be appreciated.

BMW

R69

The R69 was an important engine for BMW because it led to the development of the R90. It also showed that BMW was starting to refine the visual appeal of its engines.

☙ ~ ❧

I began thinking about the exhibit and where to hold it. I thought about Up-N-Smoke, but they didn't have room for a sustained display. My next choice was the Barber Museum. That would be fitting because it was the museum that propelled the project from the beginning. But how was I going to arrange something like this? I was a much better businessman than I was a negotiator.

I contacted *Motorcycle Classics* and told them that I was going to try to arrange an exhibit at the museum during the 2007 Barber Vintage Festival. I asked them that, if I did, would they like to sponsor it? Their response was favorable, but they couldn't commit to it until they knew what form sponsorship would take. That was understandable, but I felt that step one was complete.

Christmas sales had allowed me to save enough money to travel to the Barber Museum again. I arranged to visit the museum in February to photograph more engines and present the idea for the exhibit in person. Jim owns a rare Dakota 4 motorcycle and he wanted to see if the museum was interested in buying it. We'd travel to Birmingham together again with Jim's discounted airline passes. We arrived at the museum on a cold Thursday morning and asked to speak with Jeff Ray.

Jeff came downstairs and greeted us with his professional politeness that I remembered from our first meeting. I asked if he had time to discuss a few things with us and he led us into a conference room. We discussed the project and he said that he had noticed my ads and the press releases. I always tried to include a mention of the museum in every press release.

Then Jeff suggested that a donation to the museum might be in order. Seemed reasonable to me. I got out my checkbook and asked if $100.00 would be okay. For a brief instant I saw an embarrassed smile flash across his face. Jeff simply said, "We take what we can get." As a director of a charitable organization, Jeff wasn't allowed to require a donation, so he couldn't just tell me what amount would be satisfactory. And I didn't tell him that was all the money I could afford. If he were expecting more, I'd have to make it up to the museum in the future.

With that awkward moment passed, I outlined the exhibit idea. Even though it was frigid outside I was sweating through my shirt. I didn't sound anxious, but I sure felt it. Jeff listened patiently to the proposal and about *Motorcycle Classics'* potential sponsorship. He asked some good questions and, as the discussion went on, he started to like the idea. We talked about some of spaces in the museum that would be suitable for an exhibit of thirty prints.

In the end Jeff indicated that the museum would likely be willing to host an exhibit during the festival. Of course he'd have to discuss it with the board before he could commit to it. I told him that I would write up a formal proposal and send it to him in a few weeks. Step two was complete.

Driving back to the hotel in the rental car I couldn't help thinking to myself, "My photographs in a museum ... what a great thing to happen to little Danny Peirce from little Universal City, Texas."

(Motorcycle courtesy Clay Walley)

SPITFIRE MK II

Peckerhead Clay Walley wanted me to include his BSA Spitfire engine in the project. I am a sucker for BSA engines. It took a few months before we could get the bike and me in the same place with a camera. We finally hooked up at Dean Baker's house where I photographed the BSA and Dean's Tiger 110. Clay pulled the large touring tank off of the Spitfire for a better look at the cool, domed valve cover.

I posted the Spitfire Mk II picture on my eBay store and overnight I received a note from a BSA expert who wanted to point out a flaw in the engine. He told me that the Mk II was fitted with Amal GP carburetors and the engine pictured had inferior Amal Concentric carbs. He went on to inform me that any true BSA fan would cringe at the sight of it. But, he did say that he thought it was a pretty good picture otherwise.

I thanked the expert for his note because I like to learn about each engine in the project. However, I had to point out that I was engaged in an art project about the graphic nature of motorcycle engines and wasn't producing a historical treatise. Very few of the engines in the project are absolutely factory correct. That's the way I like it.

At the next Peckerhead Friday I told Clay about the issue of the carburetors. Clay said that the expert was correct and that the bike did indeed come with Amal GP carbs. He had replaced them with the Concentrics. Then he told me the story.

Several years ago a fire at the British National Motorcycle Museum had decimated a good portion of its collection. One of the bikes destroyed was the Triumph streamliner known as the Texas Ceegar. This was the streamliner that set land speed records on the Bonneville Salt Flats in the 1950s. It was the motorbike that gave the Triumph T-120 its Bonneville name. After the fire all that was left of the streamliner was a semi-melted frame and a charred engine.

The Texas Ceegar was originally built in a shop in Fort Worth, Texas. When the NTNOA found out that the streamliner was ruined, they asked the museum to send them the carcass so they could attempt a restoration. Ed Mabry still had the original molds for the fiberglass streamliner body stored in the attic of his shop.

The Norton club rounded up a group of talented and dedicated volunteers ... many of them Peckerheads. Ed Mabry spent months painstakingly straightening the frame and intricate steering system. Simultaneously, work was being carried out on the body, the engine and painting. After nearly two years the restoration was almost complete. The engine in Texas Ceegar had sported the quality Amal GP carburetors. They would need a set to complete the engine rebuild.

Clay happened to have a set of GPs on his Spitfire. He generously donated his GPs to the project. They remain today hidden inside the fully-restored Texas Ceegar that is back on display in the British National Motorcycle Museum. Clay says that the Concentrics are better for riding around town anyway.

(Motorcycle courtesy Dean Baker)

TIGER 110

I wanted a modern style Triumph pre-unit for the project. It didn't take a lot of searching before Peckerhead Dean Baker volunteered his Tiger 110. If you've been keeping track, you may have noticed that the Peckerheads own a pretty impressive collection of motorbikes. Most members have several bikes in their garage. I'm an exception with only one ... although, I've owned as many as three Hondas at one time.

All of the British engines seem purposefully and attractively designed. For me, the Tiger 110 has a civilized look to it. The kind of motorbike a gentleman would ride. Everything in its place, with a symmetrical standard that assured the rider that he was traveling on a quality machine. That's what a gentleman would see. There were plenty of hooligans, however, who saw a civilized engine waiting to be corrupted.

☙ ~ ❧

Jim took up the challenge of helping me make the frame modification successful. He insisted that we bring the Landshark to his garage to work on it. It made sense because his garage had the tools and welding equipment we'd needed, and was air-conditioned. We trailered the bike to his house, about 10 miles from my home. It was a nice offer, but I was uncomfortable with the thought of my bike taking up space in Jim's garage for who-knows-how-long.

We had a design in mind for the modification, but we had to wait until we had the sections cut out before we settled on a solution. Once we removed all the parts and wires in our way, we found that we could cut out the frame sections as one piece because they were connected with a crossmember. Measurements were made, marks were drawn, and we cut. Now I was committed.

With the frame sections removed I could get the valve cover off to inspect the valves. Removing the cover was like driving up on a terrible traffic accident. The first thing I saw was aluminum shavings everywhere. It didn't take a lot of investigating to find the real problem. A bolt that held the right cam tower had stripped out. There are two cam towers that lie across the engine head and hold the camshaft in place.

The stripped bolt on the right cam tower allowed it to lift up slightly with every beat of the engine. The camshaft had chewed itself up and was completely ruined on the right side. In addition, the lifting of the cam tower allowed oil to get underneath and out a machining port in the head. That caused the oil belch on my pant leg. It also starved the right side of the camshaft for oil and galled the lobes and bearings.

Luckily, I had a parts engine so I had another set of cam towers and a camshaft. Oh crap! I realized that I had given the camshaft to Randel to make a lamp out of. I crossed my fingers and hoped I could get it back in one piece. While I wrestled with correcting my mechanical woes, Jim was concentrating on the frame mod.

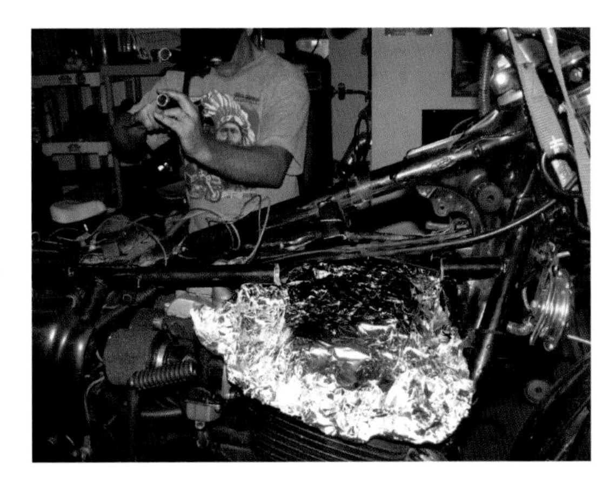

Jim measures the dimensions of the tubes after they were cut out of the frame of the Landshark.

(Motorcycle courtesy Barber Vintage Motorsports Museum)

400F

It seems that Honda fell in love with the in-line four engine during the 1970s. It produced them in a variety of sizes and packages. The 400 Four was one of those packages, and to make it special Honda gave it a signature 4-into-1 exhaust.

There is no doubt that those sweeping headers defined this engine. The motorcycle as a whole received only mediocre reviews, but it sold well because the pipes set it apart from everything else on the road. Honda used mechanical esthetics to sell the 400F. It turned a mechanical necessity into a styling statement. The visionary spirit of Soichiro Honda lived on.

~

Fortunately, my salvage camshaft hadn't been made into a table lamp yet. That was only because Randel hadn't figured out how to bore a hole through the length of it. I got the camshaft back from Randel and dug around in my parts box and managed to find all the other parts and gaskets I'd need.

Meanwhile Jim had decided that short pieces of steel rod could be welded into the ends of the frame sections. One end of the insert could be machined as a tab to mate with a corresponding tab welded in the frame. Two bolts in each of the four tab unions would secure everything solidly. Jim described his plans to Steve Ledbetter who made a detailed CAD drawing for approval. Then Steve machined the steel pieces for me. I've always been impressed with skilled machinists. Steve is modest about his skills, but the pieces he made for me showed that he had nothing to be modest about. He spent a lot of time making the eight tab pieces we needed. Steve allowed me pay for his hardware expenses, but he only wanted one of my prints in return for his time.

Jim, as well, spent hours welding the pieces in place. In the end his design worked perfectly. The sections were removable, but when bolted into place they made the frame stronger than it was originally. John Buckles, who also owns a CB750 SOHC, came over and helped me reassemble the cam towers and camshaft. I think John really just wanted to see if all this was actually going to work. After a few helicoils we were all set. The Landshark roared back to life after four weeks of work. I'll take a lot of friends over a lot of money any day.

(Motorcycle courtesy Barber Vintage Motorsports Museum)

YALE MODEL 25

I don't know much about Yales, but I know what I like. In 1912 many motorbike manufacturers were using twin cylinder engines with the same basic configuration. If everyone is using the same technology, how do you make your motorbike stand out? The answer, of course, is to give it some style. The horizontal fins on the Model 25 set it apart from everything else in its day.

ఆ ~ ౖ

I needed a book called, 'How to do things you've never done before.' I had to write a formal proposal for the museum to host the exhibit, and one for sponsorship from *Motorcycle Classics*. I was sure that there must be an official way to write a proposal but I didn't know what it was. My philosophy is: do something, even if it's wrong. I'd present the most detailed proposal I could and we'd have to start negotiating from there. The proposal would just be the starting point.

I envisioned an exhibit of thirty prints – twenty-five 16in x 20in and five 24in x 30in. The museum had concrete walls, so I would hang the prints on eight freestanding 2ft x 7ft black grid wall stands. The frames would be a consistent satin black, with no glass front so I could show off the Metallic finish of the prints.

This would be strictly an art exhibit. No pricing or order taking. I would have a table near the display with informational materials on the project and for *Motorcycle Classics* magazine. We'd have a one-hour reception for the exhibit on the Saturday afternoon of the Barber Vintage Festival.

Only three things were needed from the magazine's sponsorship. They would provide advance publicity, a few banners to promote the exhibit during the festival, and $100.00 for refreshments during the reception. The magazine could take subscription orders and sell a commemorative poster during the reception.

The proposal for the museum was almost identical except all I needed from them was space for the exhibit and access to the museum to set it up. I also asked the museum for a table, a public address system and some lighting, but I would work without them if they weren't available.

For my part, I would pay for all of the expenses for the framing and the display, and I would do all the set-up and take-down work. This would cost me a relative fortune, but I would have sold one of my kidneys to have my art exhibited in the Barber Museum.

By the beginning of April I was able to send proposals to the magazine and the museum. The plan would either work or it wouldn't. I had done the best I knew how. But I was in way over my head.

(Motorcycle courtesy Barber Vintage Motorsports Museum)

MV AGUSTA AMERICA 750

The America 750 reminds me of a body builder's physique – all massive lean muscle without an ounce of fat. This engine was designed and built to race, too bad so many of them sit idle in museums or private collections.

☙ ~ ❧

"When is this engine project going to put money into our own pockets?" Pam asked with an honest curiosity. She wasn't upset, but it had been almost two years since we invested our personal money in it. Pam just wanted to know when to expect a return on her investment. I was still putting every dollar back into the project's expenses. I knew with the exhibit coming up that it would take everything I had to make it happen. "Just give me until the end of the year," I said ... optimistically.

My main concern was the cost of framing. With thirty prints, even at $20.00 each, it would cost me $600.00. And there was no way I could get them framed for $20.00 a piece. I'd have to do all the framing myself. I figured that in my limited spare time it would take me months to complete. I would need to order frames as soon as possible. But not before the proposals were approved.

☙ ~ ❧

After several emails and a couple of phone calls, Andrew Perkins of *Motorcycle Classics* and I had fine-tuned the sponsorship proposal to mutual satisfaction. Andrew was a business-savvy guy and I liked all of his suggestions. He told me that even if the museum didn't host the exhibit, they would host it in their expo area during the festival. We also arranged for the magazine to start selling some of my prints.

It was mid-May and I hadn't heard back from the museum yet. I sent Jeff a couple of emails and called and left phone messages twice, but I hadn't heard back. In the second phone message I left I asked for a decision on the proposal by the end of May.

I found a frame supplier who could give me the frames for $480.00. They would be wooden frames painted satin black with no glass. I went ahead and bought them even though plans hadn't been finalized yet. That was the single largest expense in Trick Photography's short history.

☙ ~ ❧

Also in May, I attended the British Motorcycle Owners Association New Ulm Rally. The BMOA is equivalent to the NTNOA, and the New Ulm Rally was its equivalent to the Lake O' The Pines Rally. New Ulm is a tiny rural town in central Texas, just west of Houston. Once a year it is invaded by a hoard of British and European motorcycle fanatics. I'd been invited several times, but this was my first trip there.

Even though the Landshark was running it didn't get to go with me because I was hauling my print display instead. Plenty of NTNOA members were there so I already knew a lot of faces. The weather was mild, folks were friendly, the beer was cold and a good time was had by all. I stayed in a hotel, but people who camped at the rally told me stories about the campgrounds at night. There was something about a bottle rocket war between two sides of the campgrounds in the middle of the night. Rambunctious camping behavior must be a trait of British bike owners. Maybe I'll pitch a tent next year ... somewhere out of the line of fire.

Scott
RZ 1207

SCOTT SPRINT SPECIAL

Since this book isn't about the technical specs of the engines I won't go into detail about the Scott. However, I will recommend that you look up the information on this motorcycle. At first glance it looks like some sort of cobbled together contraption. In reality, it was the most technically-advanced 2-stroke engine for many years. Notably, it had oil injection and liquid cooling. Look it up.

଼ ~ ଼

Thursday, May 31. I came home from work and sat down at the computer in our living room to check my email. "Hey Pam, I got something from the museum." Pam was on the couch watching television. I opened the email and read. "Crap!"

"What's wrong, honey?" Pam got up and came over to read the email. "Oh honey, I'm so sorry," Pam said, as she hugged me. As she hugged me, all I could do was stare at the computer screen and read the email over and over.

Hello Daniel,

Sorry for the delay in responding to your request. I had to go back and retrieve your emails from my spam filter file. Sorry again!

After considering of your request and taking into account all the activities we are conducting in the museum that weekend (Vintage Festival), it has been determined that the museum will not be in a position to provide a 'gallery space' for the exhibit in the museum. A space in the vendor area (not swap meet) may be suitable, but will require a secure tent and such.

As you know, the understanding we have on the use of the images is based on a step-by-step approach. Before this grows to anything larger than our current understanding, we will need to enter into a more detailed, specific 'License Agreement' for the use of images and logo.

Again, sorry for the delay in responding.

Sincerely,

Jeff Ray
Executive Director
The Barber Vintage Motorsports Museum

The message was simple enough. The museum would be too busy for me to set up a display of prints. That was disappointing, but as a devout pragmatist I never count on anything until it actually happens. What put the rock in my gut was the tone of the email. I thought they liked me.

This didn't seem to be the kind of email that was sent to someone you have a good relationship with. My mind was trying to make sense of it all. Maybe I was reading too much into it. Maybe I confused politeness with respect. Maybe my donation wasn't big enough. Maybe they thought I was dishonest and trying to cheat them. Maybe they thought I was making a lot of money. Maybe they thought I was a nuisance. Maybe they thought I was a nobody trying to masquerade as a somebody.

Maybe I really was a nobody trying to masquerade as a somebody. I didn't know what it was, or what I had done to sour things between us. I felt sick, and I cursed my naivety.

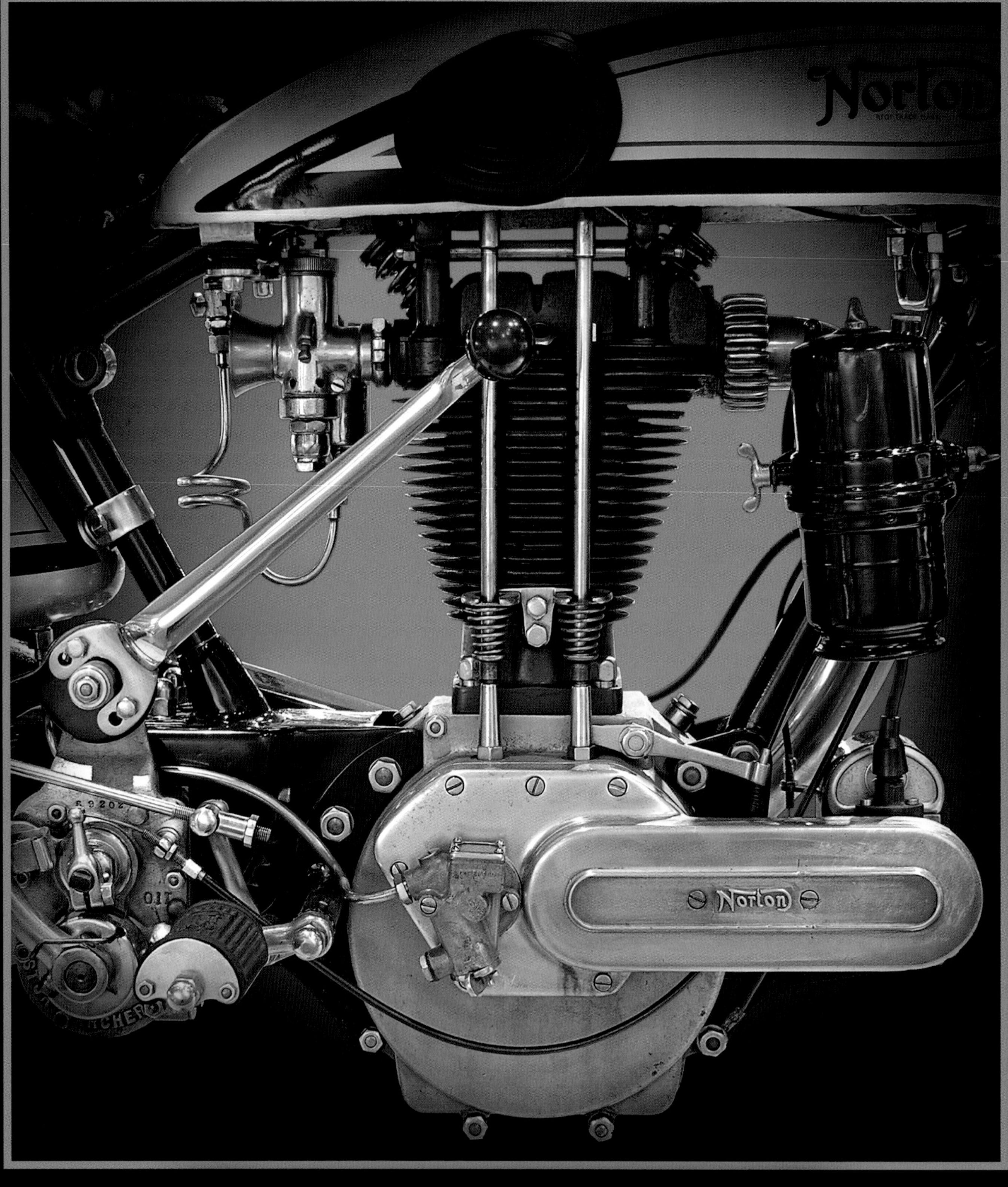

(Motorcycle courtesy Barber Vintage Motorsports Museum)

NORTON MODEL 18

The Model 18 is an attractive and well-designed power plant. So well designed, that Norton produced it in one form or another for half a century.

☙ ~ ❧

In 2001, my brother-in-law, William, and I decided to produce a joke web site. The original idea was to create a fake and humorous religion and maybe sell some coffee mugs and tee shirts. That was the original beer-inspired idea. What we ended up with was something quite different.

Combining the words 'apathy' and 'theology' we came up with 'Apathyology.' William would be the web master and I would write the content. We set up as Apathyology.com. William, being the slightly paranoid type, insisted that I not use my real name on the site. I came up with the pseudonym, Uncle Cosmo. (I really did have an uncle named Cosmo.)

It was during the writing process that things started to change. Instead of creating a joke religion, I realized that I was actually writing out my own warped philosophy for living. The name Apathyology was silly, and the content was written with a certain amount of tongue-in-cheek, but I was serious about what I was saying.

Apathyology became the art and science of selective caring. It said that there are things that we should not care about, such as greed, hate, revenge, jealousy, hurting others, etc. Caring about those emotions weakens our self-esteem and is counterproductive. There are also things that we should care about, such as life, love, family, friends, helping, giving, etc. We should care about the things we should care about, but not too much. Caring too much leads to caring about the emotions that shouldn't be cared about. In its essence, Apathyology is a healthy self-esteem emulator.

There's more to it, but that's the core of the philosophy. And that's how I live my life. I don't want to care about the negative emotions that would cause suffering to myself or anyone else.

After the email from Jeff I thought a lot about Apathyology. I couldn't help the disappointment, but I could control how I reacted to it. I knew that the people at the museum were actually very nice, and I wasn't going to let my disappointed change how I felt about them. The Barber Museum was still the best motorcycle museum I've ever seen. And I still felt an obligation to it for allowing me permission to photograph some of the engines in the collection.

William and I kept the Apathyology web site up for two years before we agreed that we couldn't afford it any longer. We never did offer a mug or tee shirt for sale. The web site content was eventually picked up and re-posted as Apathyology.info by The Universal Church Triumphant of the Apathetic Agnostic.

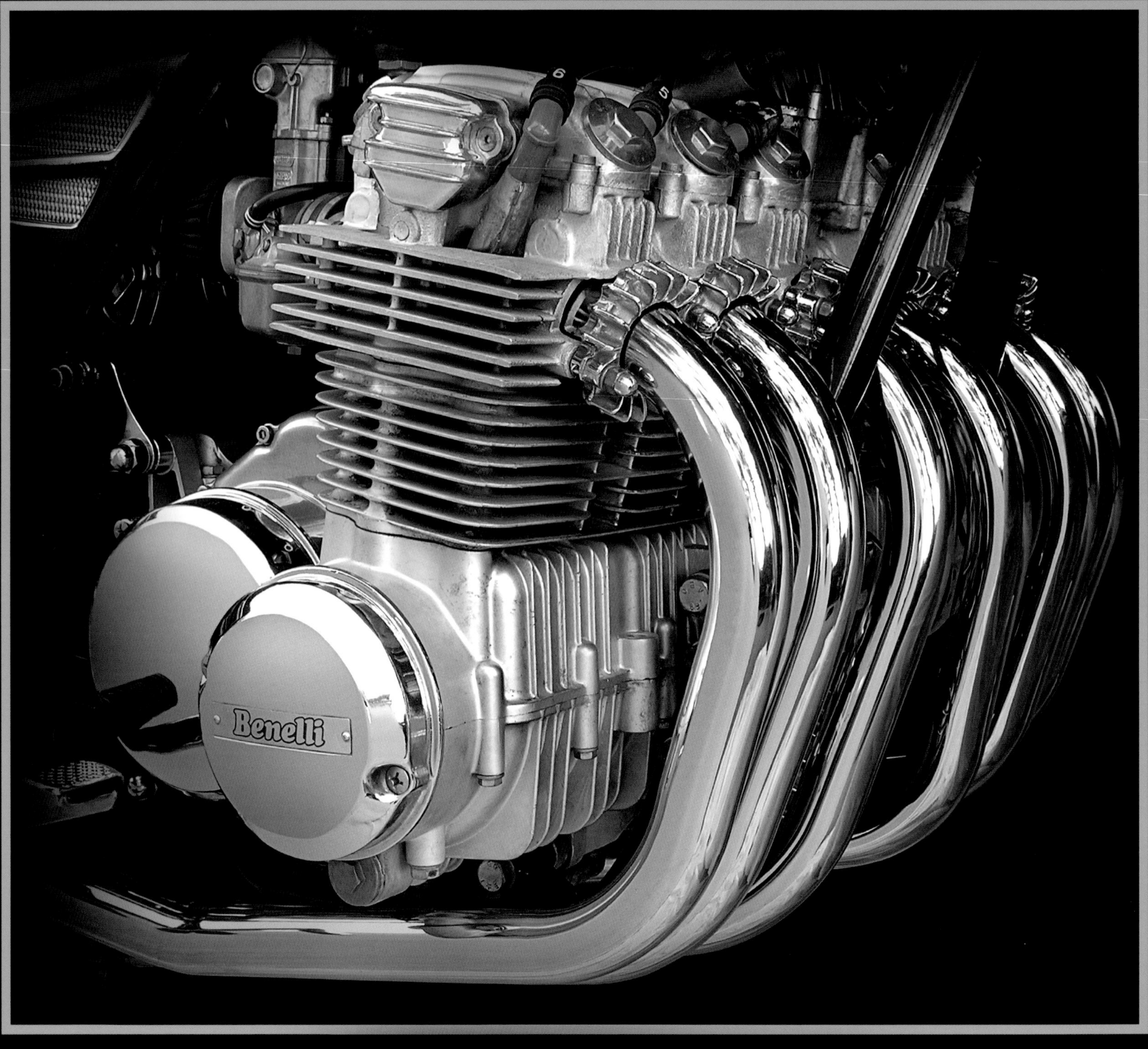

(Motorcycle courtesy Phil Dansby)

900 SEI

The Benelli Sei doesn't get the respect it deserves. Detractors always thought of it as a mutated Honda. Perhaps Benelli was influenced by Honda, but the Sei came out years before the CBX.

This is another bike from Phil Dansby's collection. It doesn't get ridden too often, but it is a runner. Phil found out that Dennis Manning of Bub Racing once made a sexy 6-into-6 exhaust system for the 900 Sei. Phil called them up to see if he could buy a set. They told him that they had the mufflers, but not the header pipes. And they weren't sure where the bending jigs were anymore.

Phil was not to be denied. As a bribe he boxed up a batch of his Texas barbecue and shipped it off to Bub Racing in California. It must have been some tasty barbecue because they built the pipes for him. A motorbike with three mufflers on each side tends to get noticed. And the sound they make resonates like a can full of very large, and very angry bumblebees.

଼ ~ ଼

The day after I received the email from the museum I contacted Andrew at *Motorcycle Classics*. I told him the museum had declined to host the exhibit. I also mentioned that I wasn't sure if I was still in good standing with them. The exhibit not being in the museum changed the entire aspect of the event, and I offered the magazine the opportunity to opt out if they wanted. Andrew just responded by saying, "Don't worry Dan, we'll make it happen."

His support meant a lot to me. My confidence was a little shaken. The exhibit would now be an outdoor event at the magazine's expo area at the festival. We agreed that, given the step down in venue, the reception wouldn't be appropriate. An outdoor photographic print display would be a poor way to show off the engine project. Nevertheless, I had already spent the money on the frames and grid wall stands, so I figured I might as well carry it through.

My daughters, Elizabeth and Catherine, had grown up seeing their dad's photographs in magazines, catalogs, books, etc, so they never thought my work was anything special. It was just what dad did. But the prospect of seeing their father's art in a museum had impressed them. They were both trying to make travel arrangements to attend the festival. I had to call them and ask them not to bother. Without the museum I couldn't see the exhibit worth their efforts to attend. It took me about two weeks to bring myself to make those phone calls.

The Benelli 900 Sei – hardly a copy of a Honda.

(Motorcycle courtesy Phil Dansby)

TIGER CUB

The Triumph Tiger Cub is one of those unusual motorcycle engines that has two attractive sides. In the end, I couldn't ignore the delicate bend of the high pipe.

I refer to this Cub as Phil's drag bike. It belongs to Phil Dansby, and he actually won a dirt drag race with it at the New Ulm Rally. I think one of Mr Budweiser's brothers talked him into it.

☙ ~ ❧

There's an old saying among folks who like to work on vintage motorcycles – "Beware the free bike." Meaning that fixing up a free bike may end up costing more than buying a new one. In the spring of 2007, I picked up a free 1980 Honda CX500D. This was my second free bike, but at least this one came with a clear title.

There are two types of people who work on old bikes. First, there are the folks who endeavor to 'restore' the bike to its showroom condition. I admire those builders because it takes a lot of time, patience and money to do it right. Since I'm usually short on time, patience and money, I'm the second type of builder. I 'rebuild' old motorcycles, which means getting them into running condition and decorating them any way I like.

The CX500 had been sitting in a barn for several years. And it wasn't running when it was put there. The deal to get it for free involved me dragging it out of the barn and hauling it away. The bike was rough and the years of Texas summer heat had taken their toll. Faded paint, rotted tires, everything rubber or plastic needed replacing, and a mixture of oil and engine coolant poured out when I removed the oil drain plug. Beware the free bike.

I had rebuilt another CX500 several years before, so I knew all the areas that would need special attention. Just for something different I decided to turn this CX500 into a stripped-down bobber. The engine configuration would make a very curious looking bob job – exactly what I wanted. Once I removed all the body parts I found a beefy and heavy-duty frame underneath. I actually started to think the bike might look really good as a bobber.

The new obsession was begun in earnest in June. Just what I needed while producing a photo project, framing prints for an exhibit, selling prints, and trying to become famous. All this while going to work every day, and doing everything else I needed to do as a husband and a father. The Lord loves a busy idiot.

The free CX500, a diamond in the rough. Very rough.

(Motorcycle courtesy Perry Bushong)

500 OHC

The Jawa 500 OHC was Czechoslovakia's prestige motorcycle in the 1950s. It was also a Swiss army knife of motorbikes. Side cars, racing, touring, it did everything it was asked to do – except sell well. Magnesium engine case with magnesium engine covers ... cool.

☙ ~ ❧

In July, I was thumbing through the latest issue of *Motorcycle Product News* and I spotted a blurb about a new Moto Guzzi book. I wasn't really interested in the book, but I read about it anyway. The short article mentioned that it was a new release from Veloce Publishing in the UK.

That started me thinking about producing a book again. I had about 45 engines images in the collection and I figured I'd have the entire 50 for a book by September. Maybe now was the time to begin looking for a publisher again. And here was a publisher specializing in automotive and motorcycle books.

I looked up the Veloce Publishing web site to learn something about it. I already liked the name because I used to drive an Alfa Romeo Sprint Veloce. I noted the contact information and started thinking about how I'd write the query.

I spent a couple of days composing the query letter. I knew publishers received a lot of proposals so I wanted to tell my idea quickly and clearly. I'd also use this letter as a template for contacting other publishers in the future. When the query letter was complete I wrote it up as an email and sent it off. Publisher query number one.

The response I got back from Rod Grainger, the publisher, surprised me. He said they were "very interested," and asked for some writing samples. Apparently, Rod didn't know how this was supposed to work. He was supposed to thank me for my interest in Veloce Publishing and say they weren't accepting new projects at this time. Either that or not answer me at all. Saying things like "very interested" was just going to encourage me.

The insecure part of my brain said, "Oh crap, now what do I do?" The ambitious part of my brain told the other part to shut up and start writing. I asked Rod to give me a week and I'd send him the first three pages so he could review my writing style and the approach I was going to take for the book.

But the insecure part of my brain was still saying, "See Dan? Be careful what you ask for."

(Motorcycle courtesy Perry Bushong)

NORMA

In the 1950s, Zundapp abandoned production of its big bikes and started producing smaller, 2-stroke motorbikes and scooters. That's all I know about Zundapp and the Norma. I photographed it because in the biomechanical world the Norma is just plain cute.

☙ ~ ❧

The Landshark had been running well for six months, but I was starting to hear a quiet tapping from under the valve cover. Well, not a problem, that's what the frame modification was for. 15 minutes after I removed the fuel tank I was pulling off the valve cover.

The tapping must have been my imagination. An inspection of the camshaft showed no damage. Great, I'll just re-torque the cam tower bolts and adjust the valve clearance. While torquing the bolts I found it. Another stripped out bolt on the right cam tower. Luckily I caught this one in time.

My brother Frank was visiting on vacation and supervised my Helicoil repair. I thought about putting Helicoils in all of the bolt holes, but Frank cautioned me to keep things simple. "If it ain't broke, don't fix it," as the Texas-ism goes.

That should have been the end of the engine rebuild story, but I'm not that lucky. I reassembled the cam system (which I can now do with my eyes closed) and managed to snap off a metal stud in the process. I was only torquing it down. I don't know if I have a future as a mechanic, but I'm a damn good demolition man.

What made the broken stud even better was that Honda no longer offered them for sale. I extracted the broken stud and gave it to Steve Ledbetter. By this time Steve had taken to calling his machine shop 'Peckerhead Machine Works.' For the price of another print he made me four new threaded studs with higher-grade steel.

I replaced the studs and everything went back together without incident. Two weeks later I rode the CB750 the 200 miles to the Lake O' The Pines Rally. The Landshark never missed a beat. I declared the saga of the engine rebuild, which I started almost two years before, officially over. My advice to anyone thinking about rebuilding their engine? If it ain't broke, don't fix it.

(Motorcycle courtesy Jerry Hatfield)

165 ST

This 1953 Harley-Davidson 165 ST belongs to noted motorcycle historian and author Jerry Hatfield. He happened to be at Perry Bushong's shop while I was there photographing one Saturday morning.

The 165 is from a group of motorbikes commonly referred to as Hummers. However, as I learned, this is not a Hummer. The actual Hummer was a model that came after the 165 ST. I also learned that there are a lot of admirers and collectors of the 'Harley Hummers.'

☙ ~ ❧

When Rod at Veloce Publishing received my writing samples, he sent me an email saying that he liked my relaxed writing style. He also mentioned something about a contract.

Rod said he wanted to have the book out in time for Christmas of 2008, and he wanted to increase the engine images from 50 to 64. I agreed with the release date, but I wasn't sure about making another 14 images. I was only producing about two images a month in my spare time.

I calculated that with the exhibit coming up in October, and then the holidays, I could probably have everything completed for him by April 2008. Rod wrote back and said he wanted the book completed by December 2007. I told him that I might be able to get the book written by December, but not with the 14 extra images. After several exchanges back and forth we settled on a deadline of January 15, with all 64 engine images.

I was already a little panicked and it was still only August. Doing the math told me that I would have to work on the book with uncommon diligence. What had I gotten myself into? I knew the answer – exactly what I asked for.

The contract arrived in September. The publishing company cautions their authors about not making their deadline. As soon as Veloce receives the signed contract they begin promoting the book. If they promote a book for six months and the author doesn't deliver, it makes them look bad in the market. They were putting their trust in me to have everything completed and to them on time. That was pretty good motivation to use uncommon diligence.

I don't take my commitments lightly. Fortunately, I've been a commercial photographer for a long time, and I never miss a deadline.

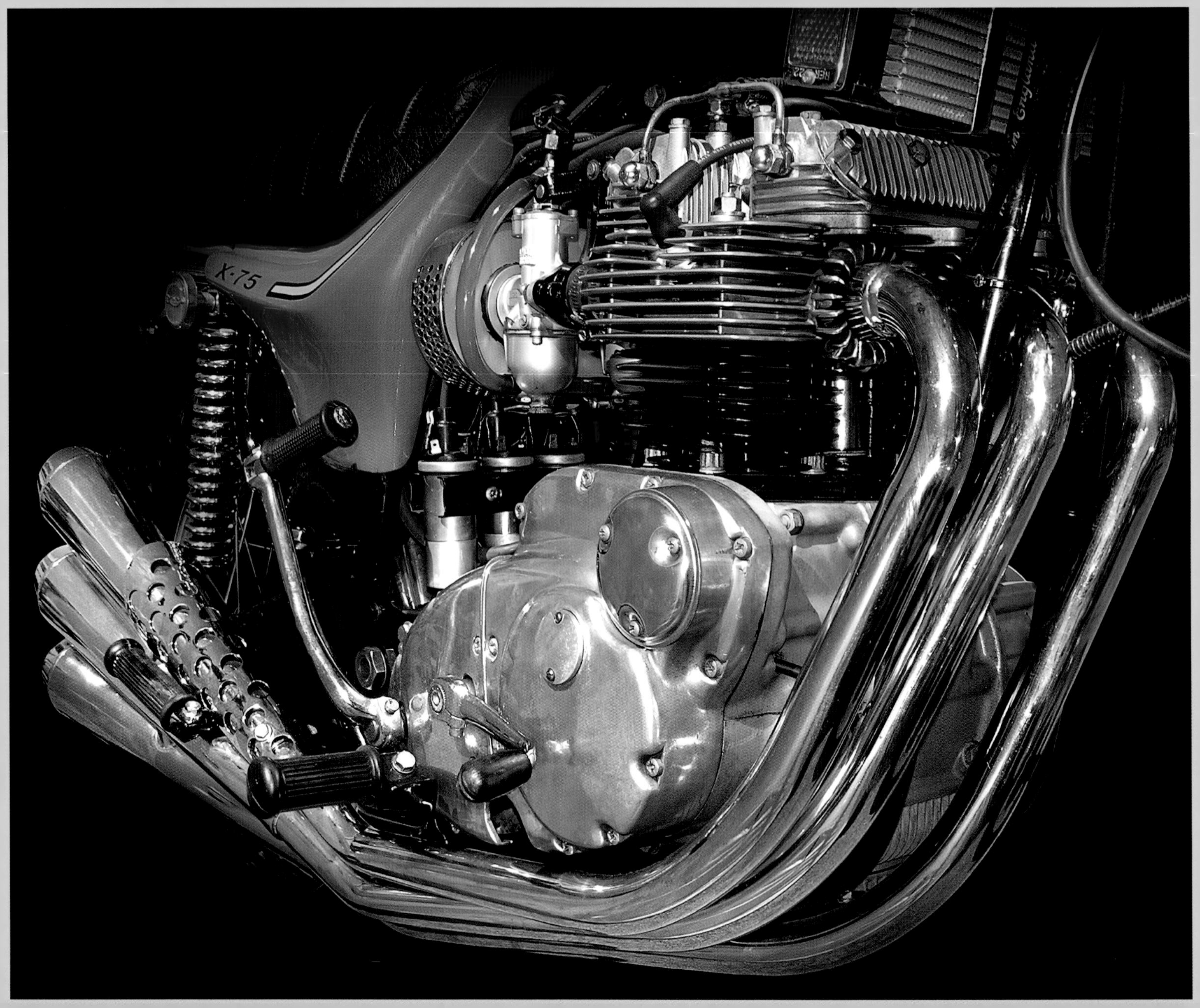
X·75

HURRICANE

In the late 1960s, Don Brown was CEO of BSA in America, and he secretly hired Craig Vetter to design a hip new BSA for the American market. Don had to do this clandestinely, because BSA as a whole was in turmoil and wasn't keen on a radically designed bike for a single market. Vetter's masterpiece was the X-75 Hurricane, and Brown eventually brought it to the market as a Triumph.

Don is a fan of the UNS Engine Project and sent me an email to see if he could purchase a print of the Hurricane. I told him that if he allowed me, I'd be honored to send him a complimentary Hurricane print. After all, without Don there would be no such thing as the X-75. It was my honor that the Motorcycle Hall of Famer appreciated my work.

☙ ~ ❧

Perry and Merry Bushong own Perry's Motorcycles and Side Cars, in Fort Worth. Perry is a noted BMW authority, especially BMWs with side cars. But Perry is also known for his knowledge of a wide range of motorcycle marques. He has restored everything from Indians to Bultacos.

It's known among local vintage bike enthusiasts that Perry also has a mighty impressive private collection of exotic motorcycles. The bikes aren't on display, and his collection isn't open to the public; but if he invites you to the second floor of his shop – take him up on it.

I needed more engines to photograph, so I asked Perry if I could shoot some of the bikes in his collection. Perry readily agreed and showed me around the second floor so I could pick out what I wanted. Perry's hands have been in quite a few of the engines in the engine project collection, and it wouldn't have been the same without his help.

☙ ~ ❧

I already had three pages of the book written, that only left me 61 more feature pages, an introduction, and a final chapter to write. Plus 14 more engine images to make. I'd have to write something every day if I could, and produce one new engine picture every week.

Dave Howe asked me if he could write the foreword for the book. Why not? Dave is the head Peckerhead after all, not to mention a better than average writer. My other choice would have been Jay Leno, but Dave asked first.

MOTO GUZZI
M 95181

FALCONE

This picture of the Moto Guzzi Falcone was actually taken early in the project. I loved the way it looked, but I wasn't familiar with it. I figured that if I didn't know what it was, then most other people wouldn't either. I figured incorrectly.

People kept asking me if I had a picture of "that horizontal" Moto Guzzi. A little research and I realized that I already had a picture of one waiting to be made into the project format. My research also taught me that this engine design powered Moto Guzzis for many, many years.

☙ ~ ❧

It was September, and the email from the Barber Museum still bothered me. I felt as if my character was suspect. I was at a loss to figure out how I screwed up our relationship. I wasn't even sure anymore that we had a relationship in the first place. They had liked the prints, and even bought some from me to sell in their gift shop. Something had changed. If this was a money issue with them, then I should demonstrate my integrity and make the first offer for a license agreement.

I just didn't exactly know what they wanted. For that matter, I'd never even seen a license agreement. I knew basically how one worked, but not every aspect. I didn't have any use for their logo so I did my best to make a sincere offer for image use. I wrote up a proposal, put it in an envelope and sent it off in the post.

Pam frowned and told me I should have had a lawyer look it over before I sent it. I didn't have a lawyer. I didn't even know a lawyer. In my mind, a lawyer, and the money it would cost to hire one, should be saved for when there's a substantial amount of money involved. Nobody was going to get rich selling pictures of motorcycle engines. I don't know if my proposal did any good, but at least it made me feel better.

☙ ~ ❧

With the exhibit coming up and production of the book beginning, I had to face the fact that there would be no time for the CX500 project. I had already replaced the wheel and neck bearings, put on new tires, and rebuilt the forks and brakes. Work on the engine had just begun when I finally put a stop to everything. I kissed the CX500 good night, covered it up, and told it I'd see it again in February.

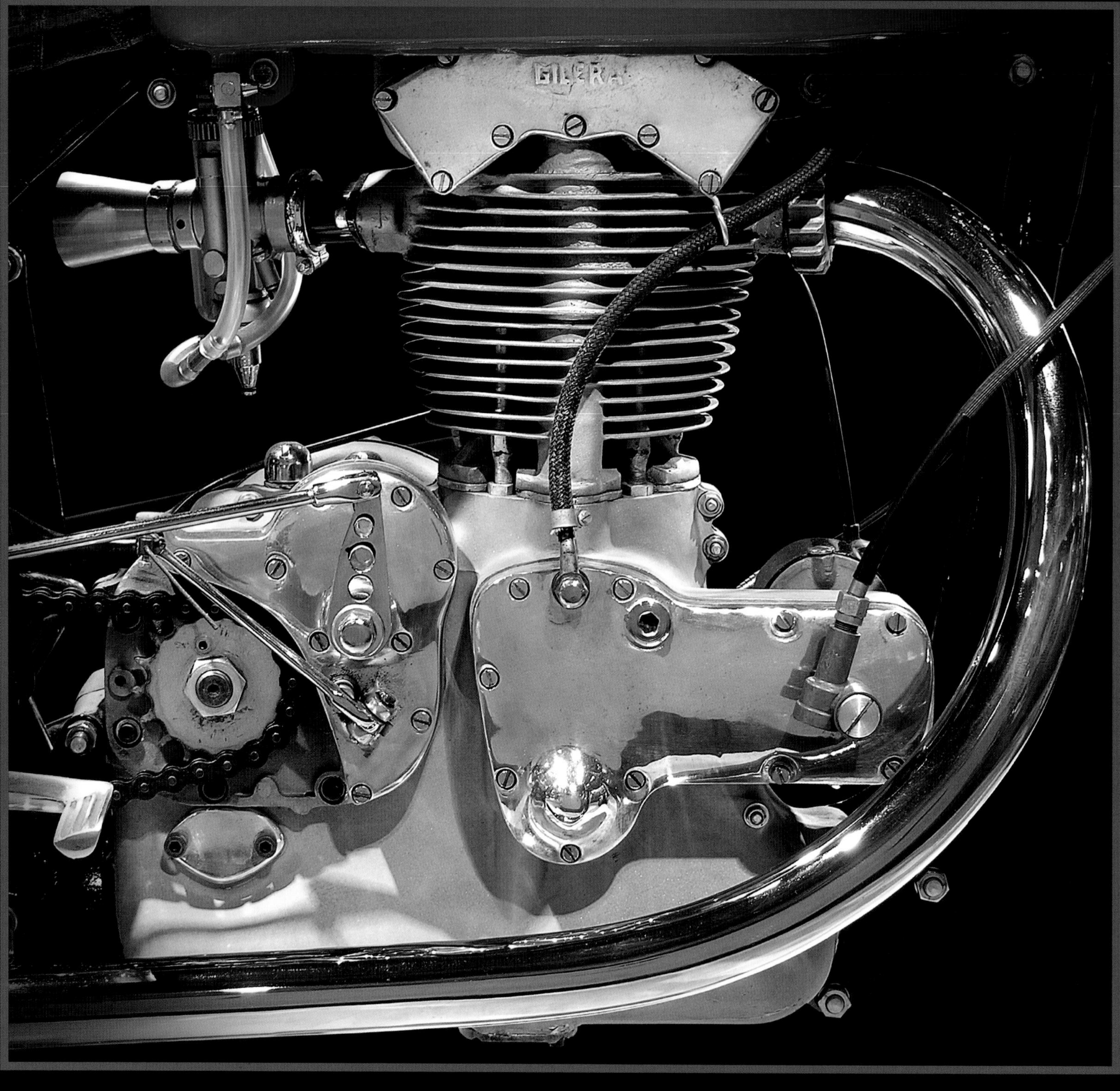

(Motorcycle courtesy Barber Vintage Motorsports Museum)

SATURNO SAN REMO

The San Remo (Sanremo) is the racing version of the Gilera Saturno. The Saturno was one of the stars of Italy's golden age of post WWII motorcycle racing. In fact, it was the arch nemesis of the Moto Guzzi Falcone. For many European races at the time, it was only a question of which marque would take the flag.

ᏣᏕ ~ ᏇᏁ

To many who haven't had to do it, framing prints doesn't seem all that time consuming. However, framing 30 large prints does take time. The prints have to be mounted on thick board, then placed in the frame and secured and then a backing has to be applied. All this may take 30 minutes a print to accomplish. That meant I faced 15 hours of framing – in my spare time. And I didn't have much spare time.

Regardless, I managed to finish the framing before the end of September. That was important because the Lake O' The Pines Rally was coming up, and the Barber Vintage Festival was only two weeks after that. October would be a critical month for me, with those two events coupled with the relentless struggle to write pages for the book.

I bought a 10ft x 20ft white canopy to cover the exhibit at the festival. It had a white fabric, which was vital to provide shade for the photos without giving a color cast. I had to devise plans of action for the contingencies of rain or high winds. This wasn't the exhibit I had hoped for, but it was the best I could do. Obsessions are merciless taskmasters.

To be certain that set-up would go smoothly, I ran a dress rehearsal in my driveway. Friends and neighbors helped me assemble everything. I'd likely have to do it all myself for the real thing. I invited all the neighbors around us to come and take a look. They knew I was doing something having to do with photographing motorcycle engines, but they had never seen any proof of it until now. Folks enjoyed the pictures at the dress rehearsal, but I realized that I'd have to cover the backs of the frames with black material to hide the cardboard I had used as backing.

ᏣᏕ ~ ᏇᏁ

I've done a lot of writing for magazines, and wasn't totally new to the discipline. However, I was accustomed to writing and rewriting multiple drafts before delivering a finished product. No time for that under the circumstances. I hand wrote a draft, typed out a draft, and then produced a final version. So, what you're reading is somewhat off the cuff. It is the result of spontaneous thought; which is always dangerous.

500

T500 TITAN

Lake O' The Pines Rally 2007. Jim and I rode out to the rally together, me on the finally resurrected Landshark, and Jim on his new retro military style Harley-Davidson. We weren't disguised as vendors this year so we were able to ride the bikes instead of hauling stuff to sell. Once again the weather was mild and the beer was cold. This was just the trip I needed to relax and mentally prepare for the exhibit and my mad dash to the book deadline. The grand marshal of the rally this year was Brian Slark, curator of the Barber Museum. The NTNOA always seems to have a knack for bringing in exceptional guests to honor. Some of the Peckerheads wanted me to speak to Brian about the situation with the exhibit. But I figured that Mr. Slark came out as a guest of the club and I wasn't going to spoil it by whining to him. Brian and his wife were genuinely nice people and I enjoyed their company during the rally.

☙ ~ ❧

Jean Akers of Biker.net was another essay contributor for the book. I asked him to write something about the Suzuki Titan. Here's what he had to say about it:

In 1967, Suzuki unleashed its superb T500 Cobra 500cc 2-stroke twin. Building a 2-stroke engine that large required that many engineering hurdles be mastered. The technology of a 4-stroke engine allows it to run considerably cooler than a 2-stroke. Since a 2-stroke produces power every stroke, there is no chance for the engine to cool, and a 2-stroke engine this big was thought to be impossible. Cooling is a major reason that a 2-stroke engine requires oil to be mixed with the fuel. This mixture helps to cool the engine and prevent seizing of the piston. Previous 2-stroke designs required the oil to be mixed with the fuel. Suzuki designed an oiling system that was separate and that could be tuned to allow the precise amount of oil to be delivered to the engine. The T500 was released as the Titan in 1968, and could out perform much larger 4-stroke engines. In a good state of tune the big twin produced 48 horsepower and 38lb/ft of torque. These numbers are comparable to a 650cc 4-stroke twin.

During the early 1970s the T500 won several major races, and this brought a lot of attention to the T500. Sales were good throughout the bike's production life. The T500 was produced until 1975 as the EPA began to focus on air pollution and the contribution made by the 2-stroke engine. The 2-stroke street motorcycle was regulated out of existence by the EPA as a major source of air pollution in 1978.

In 1970, I was drag racing my 1969 Yamaha R-3, 350cc 2-stroke twin, at a local drag strip. In those days there were two classes for motorcycles. One class was less than 750cc and the other was 750cc and over.

I beat every bike I ran against, including a couple of British twins. I knew the 2-stroke design was a power producer and my bike had been tweaked to make more power than stock. I was consistently running around 14 seconds in the quarter mile, and I had not faced a bike that could beat me. When I staged my R-3 for the next race, I glanced over to the bike I was about to race and realized it was a Titan. He had a 150cc advantage and was running a 2-stroke. I knew it was going to be difficult to beat the bike. I was hoping I could beat the rider. He beat me by a tenth of a second. That was one fast 'Suzi.'

(Motorcycle courtesy Barber Vintage Motorsports Museum)

MAMMOTH

The Munch Mammoth (Mammut) is aptly named. It is a beast of motorcycle engineering ... and I like it.

~

The plan for the 2007 Barber Vintage Festival was supposed to be similar to the previous year. Jim and Randel and I would travel there with Jim's trailer. The only difference would be that Pam was going along and I had an exhibit to set up. Shortly before we were to leave, both Jim and Randel had to bow out. That left Pam and me, and now I had to find a trailer to haul my prints and display in. I had a hitch installed on our van and I found a trailer to rent that didn't have U-Haul plastered all over it. We left at 3:00 in the morning and I drove straight through with only a couple of stops. We arrived Thursday at 4:00pm. We located our swap meet area and then went over to set up the canopy for the exhibit at the *Motorcycle Classics* area. The grounds had been improved with asphalt driveways between the swap meet rows. Even though it had rained before we got there, it was still mud free.

On Friday, Pam and I set up the exhibit. It looked sparse and lonely situated in a large open space. It didn't look bad, but it didn't look all that impressive. The kind of thing that folks would look at if they happened to walk by. We made ourselves comfortable at our swap meet space. I wasn't selling full sized prints this year, so the magazine could take orders if they wanted. We were set up to sell the photo magnets and some tiles. Of course, my goal was to sell enough to pay for the trip. But this was also Pam's vacation for the year so we weren't completely focused on making money. We invited Mr Budweiser to our camp and we spent most of our time talking to folks and watching the crowds pass by. Pam was genuinely surprised at how many people were already familiar with the project, and how many people already owned a print.

The weather was mild for autumn and the sun shone in a clear blue sky. Saturday was much like Friday, hanging around with Mr. Budweiser, talking to folks and selling a few magnets. The weather was forecast to deteriorate Sunday afternoon so we packed up the exhibit on Sunday morning and left a little earlier than planned. This was still Pam's vacation so we had made plans to stop in Shreveport, Louisiana, and stay at a casino before we drove back home. All in all Pam and I had a good time and a pretty good trip. On one hand I had accomplished my goal of having an exhibit in 2007, thanks to the generous people at *Motorcycle Classics* magazine. On the other hand, it was an exhibit by definition only. I didn't feel like an artist. I felt like an obsessive guy trying to fulfill a goal regardless of the cost or effort.

Fortunately for me I didn't have time to dwell on this because I had to continue with the production of the book. But at least now I had an exhibit that was ready to set up at a moments notice. I'd have to try again in the future.

The exhibit at the 2007 Barber Vintage Festival.

(Motorcycle subject courtesy Barber Vintage Motorsports Museum)

'58 EAGLE

If dependable is sexy, then the Cushman Eagle engine is obscene. It's the engine that launched a thousand paper routes. My brother's buddy, Greg Saunders, had one when I was a kid. The bike had a springy seat and a hand shifter, and it looked a little homemade. In my mind this was hardly a motorcycle at all. But that didn't mean I didn't want one.

~

The 2007 holiday season had slower sales than the year before. That suited me because I needed my time to work on the book instead of packaging and mailing prints. I had stopped all advertising and promotions in October to minimize new orders. I received orders anyway, but not enough to be a major distraction.

By the end of December, Pam was starting to become annoyed by the work on the book. For the past three months I'd been fairly useless around the house. And with the Christmas holiday and the deadline rapidly approaching, all I could do was juggle print packaging, writing, and keeping up with social obligations. It was difficult. I kept up with everything, but there were several compromises along the way. So, Pam being tired of my crammed production schedule was understandable. I was tired of it, too. With only weeks left to finish everything, it would be a big relief when I finally sent the manuscript to the publisher. There would still be more work to do after that, but it wouldn't be a nagging constant effort.

~

In late December I received an email from my half-brother Rob who was stationed in Afghanistan with the US Air Force. The email was subject titled, "Whoda thunkit.":

So, I'm sitting here at Camp Eggers, Afghanistan, when someone hands me a *Motorcycle Classics* magazine that someone had sent them from home. I'm flipping through the magazine, when lo and behold; I come across a fancy layout where my brother is advertising his motorcycle engine porn. I had to show all the guys in my office (many of them are bikers). Just thought I'd share that with you. Everything is going OK here. I hope that you and your family are well. Take care.

Rob

TSgt Robert F. Peirce
Chief of Travel
CSTC-A CJ8
Camp Eggers
Kabul, Afghanistan

I don't know if that meant I was famous, but it was pretty cool. I can't say if I'll ever be really famous or not, but as long as my family thinks I am, then that's good enough for me.

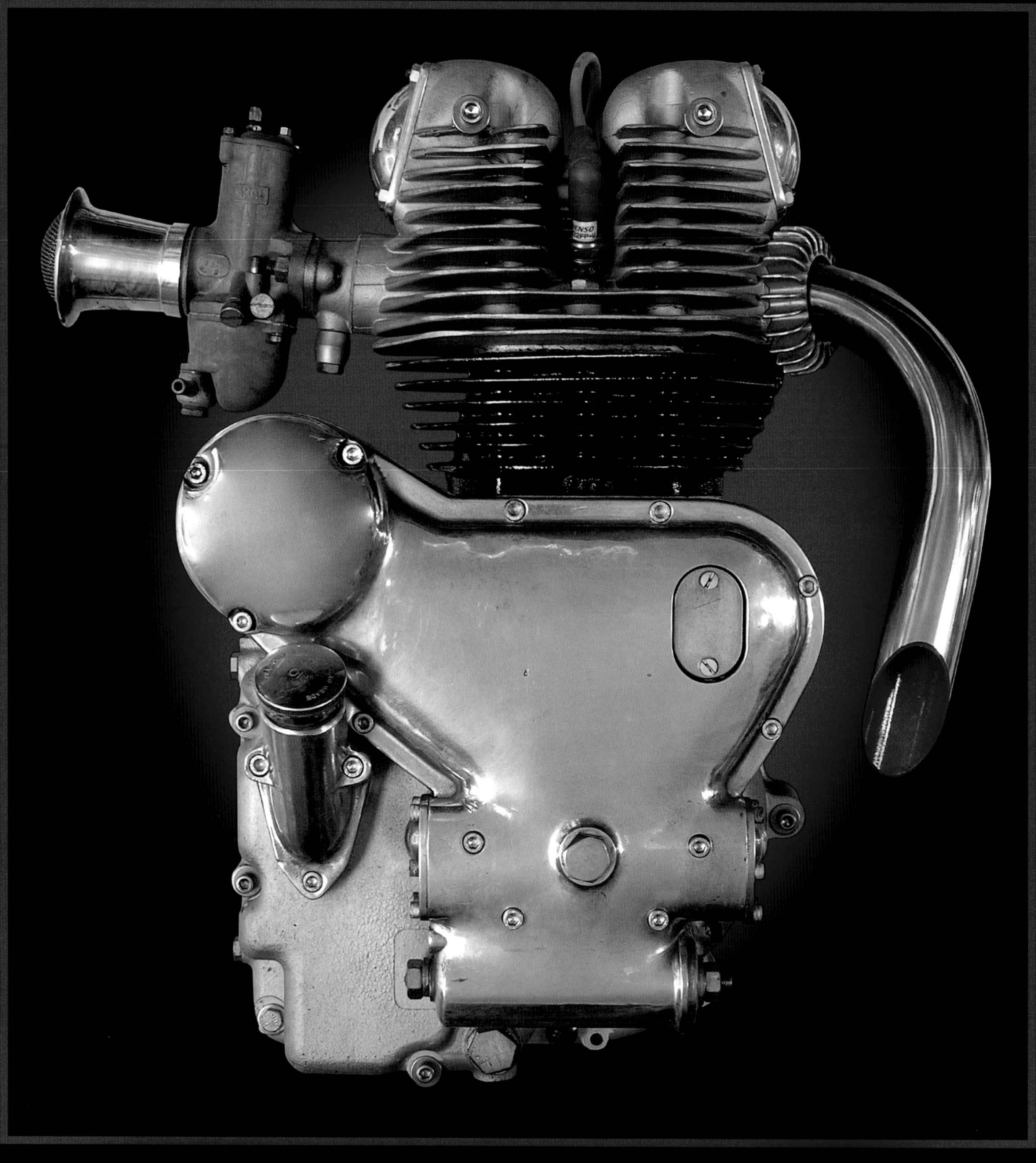

(Motorcycle courtesy Phil Dansby)

INTERCEPTOR

January 1, 2008. To end the project, and the book, I went back to the walls of Up-N-Smoke BBQ to photograph this Royal Enfield Interceptor engine. The project essentially began with the half engines that decorate the walls of the Up-N-Smoke barroom, so this is an all-together fitting way to end it.

This officially concludes the Up-N-Smoke Engine Project – at least as an obsession. I'll still add engines to the collection every now and then just because I like them. There are still a lot of sexy and attractive engines out there, but I'll take them as I find them instead of going out and looking for them. So many engines ... so little time.

I'm not making any goals for 2008. Maybe I'll make a few simple ones like finishing the CX500 bobber project, and possibly another stab at an exhibition. Pam has a long list of backlogged chores for me to do around our estate. That suits me just fine.

What I've just written makes it sound like I'm retiring. That may be what I'm thinking right now, but I know – I know – I am absolutely certain that another damn obsession is lurking just around the corner. What form it will take I cannot say right now. But as sure as the sun rises in the morning, you can bet there'll be another adventure for me soon. You can also bet that I'll be making it all up as I go along.

Until then, and even after, you'll likely find me at the Peckerhead table at Up-N-Smoke BBQ every Friday evening. If you ever find yourself in Keller, Texas on a Friday afternoon, be sure to stop in at Up-N-Smoke, and say hello to the Peckerheads. We'll greet you warmly, talk some motorcycle, and maybe even buy you a beer. There's always room for one more.

The official Peckerhead table at Up-N-Smoke. The picture in the background to the left of the door is the first print produced from the engine project.

HOW THE PICTURES WERE MADE

I don't believe that the Up-N-Smoke Engine Project will give me a reputation as a great photographer or artist. The art in this book is the engines and the artists are the engineers who created them. All I did was present the engines in an artful manner. My contribution to these pictures is more as a digital retoucher than as a photographer. Except for a small handful of images, all of the engines pictured in this book were taken wherever I could find them, and with whatever light I happened to have at the time. If I was lucky, I would be able to position the bike for the most flattering light angles. Usually, I just had to work with what I found.

The look of the engine pictures evolved as the project went along. By the time I had a dozen images made I had begun to build a routine for retouching them. However, there was never any strict standard for how each engine would look. I treated them individually, understanding that each engine had its own personality.

So, for those of you who like to know how things work, I've provided this chapter for a step-by-step tour of the production process. In the retouching steps I can only show you what I did and not how I did them. Photoshop is a complex tool, and going into that much detail would take a whole other book. I've used Photoshop almost every day for the last 12 years, and it still seems like magic.

Taking the photograph
Several of the engines were photographed in my studio. Somehow, high-end professional cameras and thousands of dollars worth of lighting equipment make a difference. For shooting all the rest of the engines on location I had to keep things light and portable. My kit consisted of a Nikon Coolpix 8700 (an 8-megapixel camera that was

obsolete six months after I bought it), a lightweight tripod, a cable release, a 3ft x 6ft folding white fabric reflector, and a white bed sheet. I could even carry all this in the Landshark's saddlebags. When I could I would try to position the bike in the shade. I strategically cut four slits in the reflector so I could set the reflector in front of the engine and poke the camera lens through. The reflector wasn't used to reflect light onto the engine; instead it was used to give the chrome and polished metal an even surface to reflect. The white linen bed sheet was spread out under the bike so the exhaust pipes would be less likely to reflect the ground. A photo session with this setup lasted 5 to 15 minutes depending on how many angle changes I made. I never used a flash on the engines.

The pictures without any retouching usually look something like this. I often take the picture with more room around it than I need, so I'll have more cropping options later on. At this point I study the picture to decide what I can and can't do with it.

Bystanders, or the bike owners, would often help me by holding the reflector while I fumbled around with everything else. I always planned my shots keeping in mind that they would be heavily retouched later. The folks watching me shoot rarely understood that and were frequently confused with why I was satisfied with a quick shot of the engine while the bike sat under a tree. Time and time again after seeing the finished product people who had watched would remark "it didn't look that way when you took the picture." The only drawback to my method of photographing engines was that I had a record of hits and misses. I found that I couldn't use every picture I took. Extreme light contrasts, unretouchable reflections, and camera shake made some engines unusable. I never did get a good picture of a Panther engine even after two attempts.

The retouching

I determined early in the project that I wanted the pictures to be a realistic and classic portrait of the engines. There can be temptations to use Photoshop filters to render the image as an artful abstract. I didn't think that the engines needed to be anything other than what they were. A couple of the pictures have small amounts of Photoshop filtering, but nothing that affects the sense of realism. Besides, I tend to look at the over use of Photoshop filters to be a lazy shortcut.

We'll use the production of the Tiger Cub as an example for the steps that each engine picture goes through.

To finish things up, I apply a colored border frame, a standard dimension black background for the print and select a title. But I think the key to doing all of this is not to do it all at once. I chop these steps up into days so I can give my next step full consideration. When I declare a picture finished is when there is simply nothing left to do to it. After all, they say the secret to producing good art is knowing when to stop.

To conclude, I am, once again, reminded that the real artful interest in these photos are the engines themselves. All my work does is put them on display. I guess, in essence, that's all that any photographer can do. We just take the art that already exists in the world and make it two-dimensional and portable.

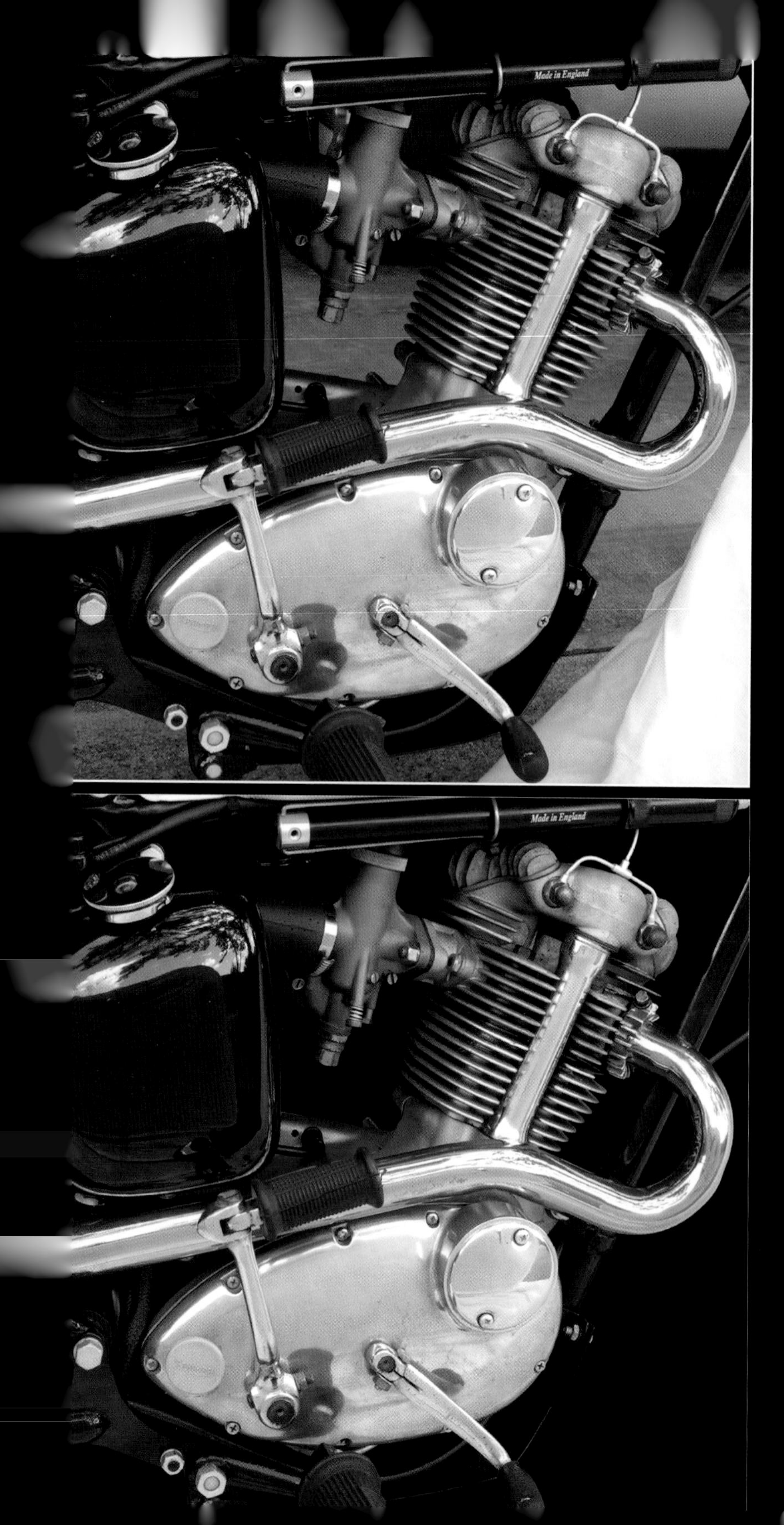

Step 1, the crop. Cropping is a simple yet critical part of the process. I toy with different crops until I find what I want. I have to decide what to leave in and what to leave out. I let the subject determine the final crop dimensions. The engines won't fit in to one pre-determined box. In the Tiger Cub I had wanted to see if I could include the muffler in the picture, but it was too far away from the main subject. I did manage to keep most of the oil tank.

Step 2, the outline. Using the Path tool, I draw an outline around all of the open spaces. This is the most time consuming part of the process. The average engine takes about an hour to outline completely.

Step 3, the background. With the outline complete, I can work on the background. I'll turn the background to black and then work on the rest of the image. Later I'll decide if I want to apply a colored glow to the background if there are too many black parts on the engine that won't show up against the black background

gives an over abundance of blue in areas that should be neutral. It's a subtle art deciding what will have color removed and what parts need to have color to look normal. When I'm finished some engines have hardly any color at all.

tep 5, clean up. Up until this point the steps are mostly technical. Clean up is the step where I get to use what little artistic skill that I have. I remove or modify any reflections that I don't want. I also smooth out metal surfaces if I think it will make a stronger picture. Every engine is different, and I have to be careful not to over etouch. Over retouching and clean up can sap all of the character out of a picture.

Step 6, vignette. Vignettes are a standard of classic ortraiture. The fading to black on the edges of the frame

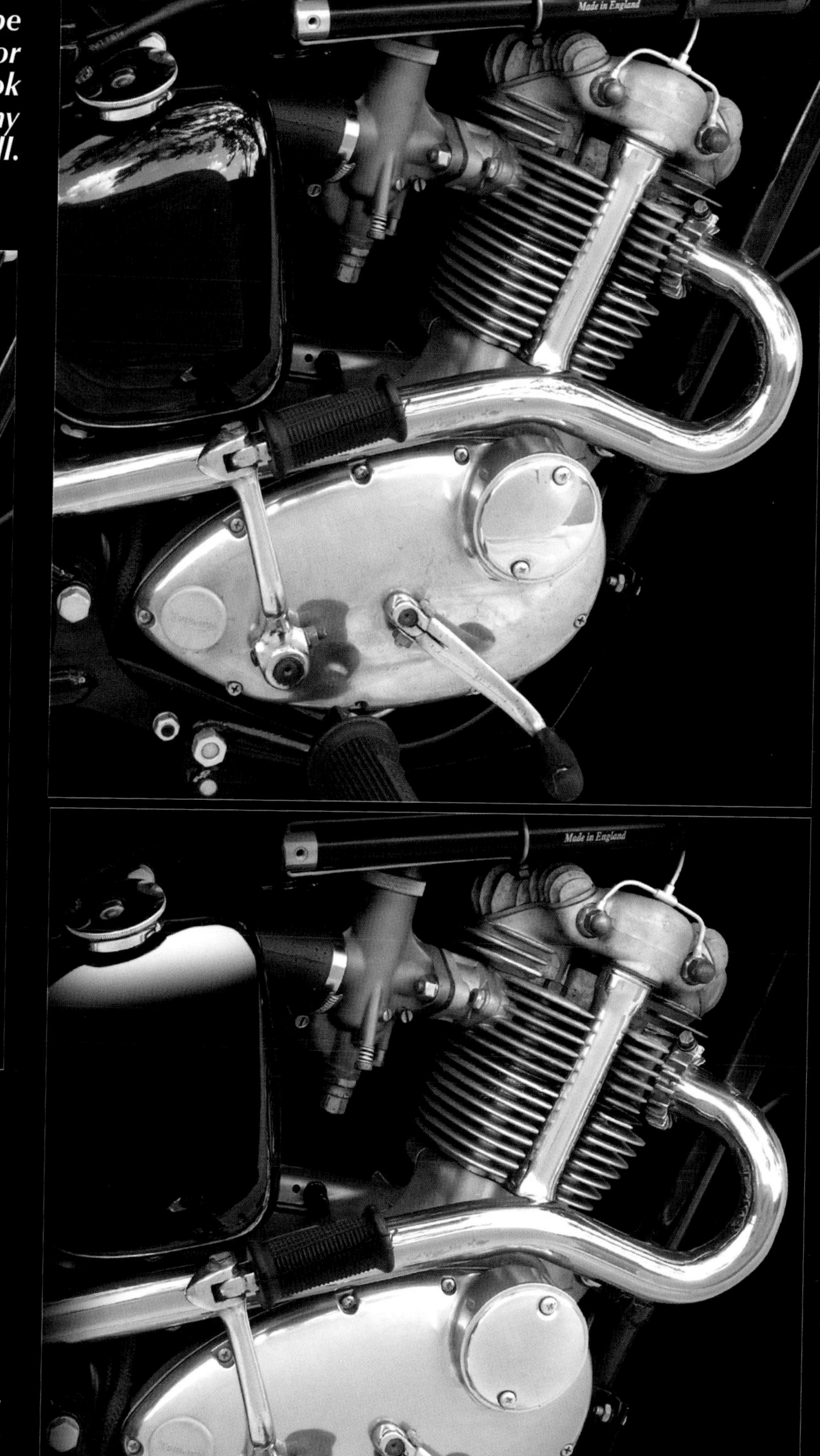

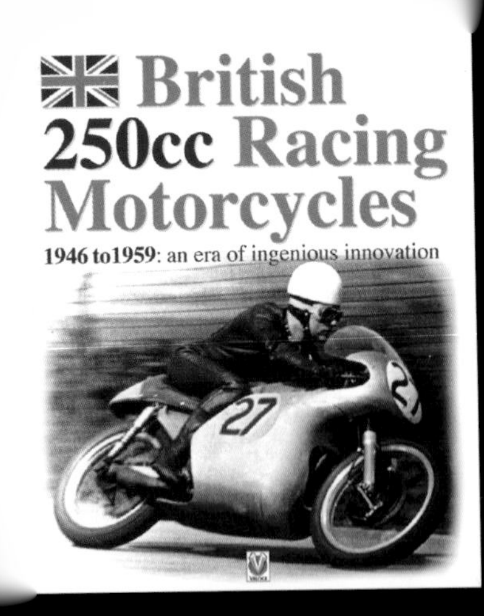

ISBN: 978-1-904788-12-6
£15.99

ISBN: 978-1-845841-59-1
£29.99

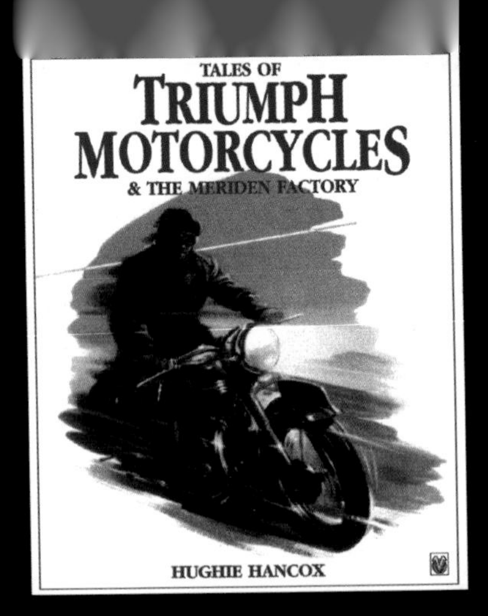

ISBN: 978-1-901295-67-2
£12.99

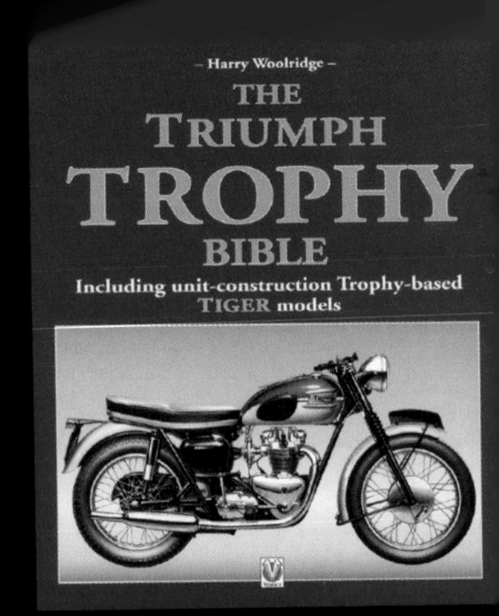

ISBN: 978-1-904788-02-
£30.00

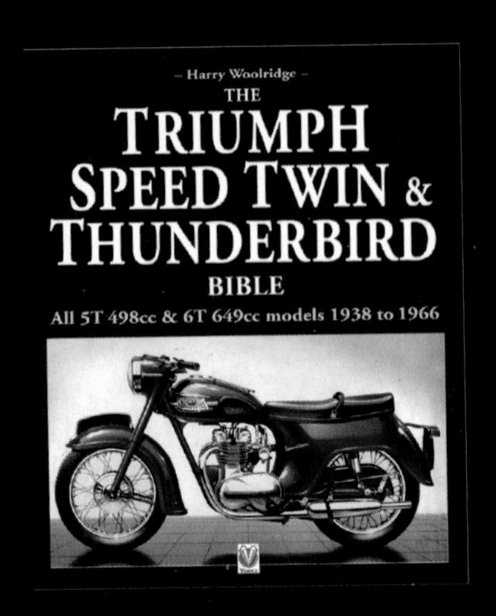

ISBN: 978-1-904788-26-3
£25.00

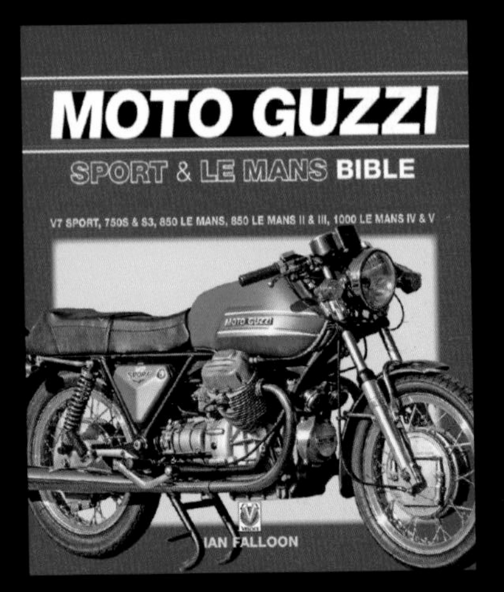

ISBN: 978-1-845840-64-8
£29.99

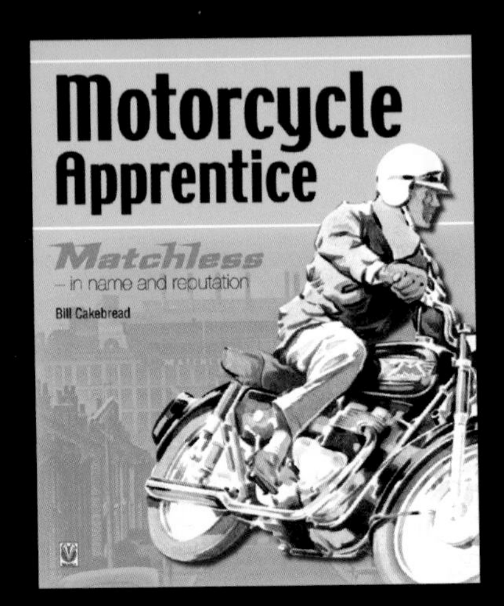

ISBN: 978-1-845841-79-9
£17.99

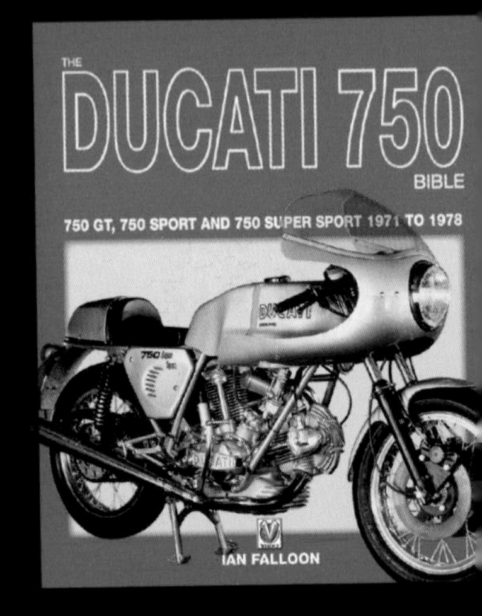

ISBN: 978-1-845840-12-
£29.99

Note: Prices subject to change, P&P extra. Email sales@veloce.co.uk or call +44 (0)1305 260068 for more details.

WWW.VELOCEBOOKS.COM

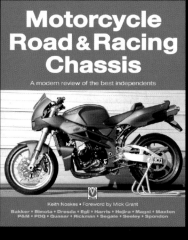

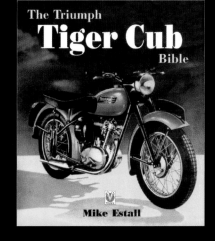

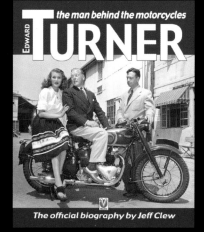

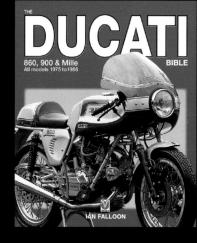

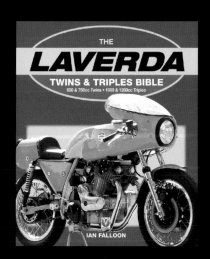

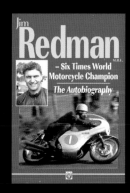

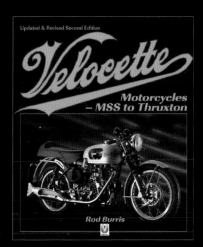

INDEX

LAVERDA